Taking Privacy Seriously

Taking Privacy Seriously

HOW TO CREATE THE RIGHTS WE NEED WHILE WE STILL HAVE SOMETHING TO PROTECT

James B. Rule

UNIVERSITY OF CALIFORNIA PRESS

University of California Press
Oakland, California

© 2024 by James B. Rule

Library of Congress Cataloging-in-Publication Data

Names: Rule, James B., 1943- author.
Title: Taking privacy seriously : how to create the rights we need while we still have something to protect / James B. Rule.
Description: Oakland, California : University of California Press, [2024] | Includes bibliographical references and index.
Identifiers: LCCN 2023053673 | ISBN 9780520382626 (hardback) | ISBN 9780520401587 (paperback) | ISBN 9780520382633 (ebook)
Subjects: LCSH: Data privacy.
Classification: LCC HD30.3815 .R854 2024 | DDC 323.44/8—dc23
 /eng/20231219
LC record available at https://lccn.loc.gov/2023053673

Manufactured in the United States of America

33 32 31 30 29 28 27 26 25 24
10 9 8 7 6 5 4 3 2 1

Contents

Preface *vii*
Acknowledgments *xxiii*

Introduction: Careering Down a Road Hardly Anyone Wants to Take *1*

1 Don't Blame Technology *28*
2 Ban Personal-Decision Systems That Violate Core Values *62*
3 Require Consent for Disclosure *89*
4 Make Personal-Data Use Minimal, Transparent, and Trackable *132*
5 Institute a Right to Resign from Personal-Decision Systems *159*
6 Create a Universal Property Right over Commercialization of Data on Oneself *195*
7 Conclusions *219*
8 The Future *261*

Appendix 1: The Eleven Reforms *271*
Appendix 2: International Privacy Affirmations vs. Privacy Setbacks, 1983–2019 *276*
Index *293*

Preface

When I came to live in the liberal city of Berkeley and moved into my modest (but expensive) new home, my first concern was to establish the essential services. Water was already flowing through the pipes, electricity through the wires, and information through internet connections. But what was I going to do with my trash? I checked the city's website and put in a call to the recycling services.

The bright-sounding young woman at the other end was welcoming. For starters, she said, "We'll need your California driver's license number."

"But why," I wondered aloud, "is this necessary?" After all, I was hardly planning to drive my garbage anywhere. I just needed to know the fee for collecting it, and where to send the check.

"Just the same," the voice responded, "the driver's license number is indispensable."

"What about people who don't drive?" I asked—then realized that I was getting myself in trouble, being added to her smartass list.

"No driver's license number," came the frosty response, "no trash collection." My resistance didn't last much longer than it takes you to read this—and probably that is just as well. Otherwise, I might now be sitting here, putting the finishing touches on this book, with accumulated trash rising around my ears.

But why was *that particular* number so important to *that particular* (relatively tiny) bureaucracy? Probably because it processes checks from its customers through a company that insures each check against nonpayment. These companies make their profits by identifying the minority of checks presented to their customers that appear suspect and declining to insure them—effectively rejecting the check. To do this, they collect and analyze information on countless consumers' financial situations—never disclosing much about their sources and methods—in order to spot those likely to go bad.

But still: why was the *driver's license number* so important? Apparently because a key source of information for such insurers is California's Department of Motor Vehicles. Provided with a driver's license number, the DMV will disclose (at least to companies like these) the name of the person to whom the license is issued. If the name on the check and the name corresponding to the license number in DMV records fail to match, the company will all but certainly reject the application.

Two decades into the twenty-first century, I should have taken it for granted that at least one set of records, somewhere, would govern anyone's access to any new service or relationship. I have in mind here not just any records, but those created by *organizations* to support decision-making in regard to the people involved. The decision-makers are agents of government or business who make their way by tracking our pasts, so as to better manage our futures. It is increasingly rare to find any area of life—from health care access to dating services—that does not rely on such scrutiny to shape treatment of those they deal with. The *ensemble* of these organizations sell and exchange personal information far and wide, seeking mutual advantage in deeper knowledge of the background and susceptibilities of each person who crosses their radar.

Surveillance and Control

Such highly focused attention to personal data is simply a manifestation of universal social processes—surveillance and control. In virtually all settings, parties (either individuals or organizations) carefully track each other's actions, trying to anticipate what the other will do next and judge how to influence those actions. Spouses do this with each other, parents with their children, neighbors with neighbors. On a larger scale, manufacturers do this with suppliers, and bosses with their staff. Analysts call the influence thus sought over others' conduct *social control*—all those processes that guide and goad people into doing what they are "supposed" to do.

For most of human history, memory has served as the main medium for tracking and recording human actions for such purposes. Today the sheer scale of institutions requires that they rely on vast computerized databases to manage their dealings with their millions of subjects. One cannot imagine a modern taxation system without detailed case records of each taxpayer's situation—of how much liability was assessed, for example, or what was paid and when. Still less conceivable would be any system of consumer credit without computerized surveillance, or an agency regulating the actions of drivers and vehicle owners, without appropriate databases. In dealings with bureaucracies that furnish key services—as much as within the nuclear family—we are known by our histories. How we can expect to be treated tomorrow depends on our records of today.

There is nothing inherently sinister or mysterious about surveillance and control. These processes are as basic to everyday social life as conflict and solidarity or cooperation and competition. Nor do we necessarily experience social control as constraining or oppressive. We may think of our passports as liberating social inventions that make travel abroad possible, for example—at least until we reflect that international travel a few centuries ago required no passports.

But sometimes these same processes take much darker forms. Authoritarian governments may seize the passports of citizens held likely to criticize the regime, either to keep them under supervision at home or to prevent their return from abroad. Surveillance systems, after all, serve the most diverse social purposes, from the allocation of medical care to the pursuit of political dissidents. At worst, those subjected to them may have to fear a knock on the door in the middle of the night.

And in some cases, the authorities don't even bother to knock: "Onree Norris was watching the news when the explosion went off inside his home. 'Something went off like a bomb in my house,' the now-81-year-old remembered from the February 8, 2018, raid on the home he has owned for half a century."[1] It was the Harris County (Georgia) Sheriff's Office Special Response Team, smashing down his door with a battering ram. The explosion Norris heard was from a flashbang grenade, which "dynamic entry" teams use to disorient their targets. But Norris wasn't the target of this "no-knock" raid. Deputies had the wrong man and the wrong house. Norris later reported that the heavily armed deputies used a battering ram to break down four doors before finding him in his hallway. The intended target was a suspected drug dealer living next door, whose house did not resemble Norris's.

Attorney Darryl Scott filed a lawsuit against the deputies in the raid. "A pizza delivery man could have delivered a pizza to the correct address," he observed. But a federal judge dismissed the case, ruling that the law officers were protected by *qualified immunity*—a legal doctrine that shields police from civil suits for most actions carried out in the course of their work.

"No-knock" raids are by no means rare in the United States. A *New York Times* investigation counted eighty-one civilian deaths and thirteen among law enforcement personnel in such invasions between 2010 and 2016.[2] Sometimes these deaths result from gunfire—

but at least one was a case of cardiac arrest, suffered by an unsuspecting resident with a less robust constitution than Norris's.

Everyone knows that getting on the wrong list can cause trouble. But the form and consequences of such trouble depend on the social sponsorship and technological reach of the list in question. All personal-data systems are human creations, after all, and their repercussions reflect the competence and interests of those who create them.

Onree Norris is Black. I know of no systematic comparison, but hair-raising stories like his seem especially likely to occur to members of less privileged groups in America. I can't help wondering whether, had Norris lived in an affluent white neighborhood, someone at the police headquarters would have spotted the "no-knock" order as an anomaly before it could be carried out. In the end, even the higher authorities in this case had to acknowledge that Norris should never have been targeted, though it appears that no one was held responsible.

We need to worry about stories like this—and not simply because such outrageous events should never happen to anyone. Of course, police surveillance goes wrong in many settings where computer databases play no role. But the *scale* of damage is apt to be much greater in computer-supported, bureaucratically organized systems like those of central interest here. Totalitarianism—the ultimate disaster for the values underlying this book—is much more difficult to accomplish in the absence of direct flows of data from the populace to the authorities. The interlocking systems of personal data exchange that track us today raise the potential extent of disaster and manipulation—not just for ourselves as individuals, but also for democracy and an open society more generally.

Why Now?

Why have massive systems for monitoring personal data burgeoned in just the last forty to fifty years? Why does recourse to such systems now appear indispensable for virtually every new transaction and relationship? For most Americans, the reflexive answer is *technology*. The rise of computing and kindred information technologies, people imagine, has released a genie bent on seizing and circulating personal information wherever it can. Like a bolt of lightning that strikes a tree and vaporizes the liquids inside, computing power seems somehow destined to evaporate our data, removing it from our control and allowing it to escape in all directions.

These images deserve reconsideration. Indeed, the idea of "technology" itself requires a bit of unpacking. It's not very helpful to think of technology simply as a series of tools, apart from the social relationships and competencies that must accompany them in practice. When we speak of aircraft technologies, for example, we mean not only the physical artifacts needed for travel by air, but also the skills and mutual understandings needed to make them work. An archaeologist might discover the tools used by the early Easter Islanders to create those stunning giant heads looking out to sea that greeted the first European visitors. But understanding the *technology* of their creation would require a broader grasp of the roles played by participants in the work (who gave the orders, for example, and who did the physical labor?) and the meanings behind it.

So, when we insist that "technology" is robbing us of our privacy, we are missing something important, implying that present-day information management capabilities somehow capture the hearts and minds of Americans and impose their own directions on our lives. The reality is less mysterious and far more concrete. The prevailing social roles of computing, including the extraordinary uses made of personal information, reflect the power positions of the parties that have sponsored its development—government institutions on the one hand, corporate interests on the other. Both the goals and the thoughtways of these resourceful institutions have shaped American computing from its early days. Think of all the ways in which computing in this country has *not* developed! If we inhabited a world where, as in some ethnographic accounts, people regarded preserving a person's image as tantamount to stealing his or her soul, the evolution of personal information technologies would have unfolded quite differently.

In fact, the *institutional forms* of today's mammoth computerized surveillance systems were already in place well before computerization. Early social welfare systems, like those in Prussia from the late nineteenth century, and in Britain and the United States from the middle of the twentieth, showed the characteristic processes of compiling personal information, and of tracking and shaping the actions of each enrollee, that define such systems. Consumer credit reporting in the United States did a crude but adequate job of tracking ordinary consumers' use of credit, using technologies no more sophisticated than telephones, typewriters, and file cabinets, until at least the 1960s. What computing has done to practices like these is vastly to decrease their cost and enhance their speed and capacity, such that their precomputer forms now evoke images of Dickensian scribes and hand ledgers. The constant here is the distinctively modern notion that large, impersonal institutions can and should track the fine detail of the lives of millions of otherwise anonymous citizens and consumers—influencing their behavior and securing their compliance through precise attention to their records.

I underline this point to support a contention basic to this book: that prevailing technologies of personal-data management could be much more privacy-friendly while still serving important social needs. Technology, in other words, is not destiny. Creative energies and cash investments that now go into separating Americans from their data, and using the newly appropriated information as the basis for activities not necessarily in those people's interests, could just as well support much more heartening practices.

Living with Mass Surveillance

The fact that such major personal outcomes depend on the contents of one's records guarantees steady streams of conflict and controversy. Some data subjects complain that decisions made on the basis of their personal information are unjust or defective. They contend that the information filed under their names is flawed or inappropriate, or that their identities have been confused with others', or that rules for decision-making have been misapplied in their cases. From allocation of consumer credit to sentencing in criminal cases, such themes recur with predictable familiarity. As a result, since the 1970s, the world's more prosperous societies have increasingly framed privacy codes to address such conflicts and tensions. At last count, more than two hundred countries have adopted some form of national privacy code—or personal data protection regulations, to use today's more standard term.

Despite much variation in the strength of these codes and in the mechanisms for their enforcement, they bear strong "family resemblances." Most of them—though not American law—create national or regional personal data protection commissions charged with upholding privacy values in the public forum and, often, investigating charges of abuse or mismanagement of personal data held in file. Nearly all such codes, America's very much included, stop short of applying the full force of their strictures to the institutions of law enforcement, espionage, homeland security, and other agencies of state coercion. And nearly all such codes rely on principles of procedural "due process" in mediating between the public and record-keeping authorities. Many of these due-process rules derive from the influential government report *Records, Computers, and the Rights of Citizens,* published in 1973 by the U.S. Department of Health, Education, and Welfare. This volume first formulated what many privacy scholars have come to regard as something akin to the scriptural basis of privacy protection, in its proposal of a federal "Code of Fair Information Practice," based on five key principles. These five Fair Information Practices, or "FIPs," stipulate:

- There must be no personal-data recordkeeping systems whose very existence is secret.
- There must be a way for an individual to find out what information about him [sic] is in a record and how it is used.
- There must be a way for an individual to prevent information about him obtained for one purpose from being used or made available for other purposes without his consent.
- There must be a way for an individual to correct or amend a record of identifiable information about him.
- Any organization creating, maintaining, using, or disseminating records of identifiable personal data must assure the reliability of the data for their intended use and must take reasonable precautions to prevent misuse of the data.[3]

These five points provide a remarkably succinct statement of good practice in personal-data decision systems. The authors were an elite panel of government and business insiders, considered to be in the avant-garde of thinking on these matters when privacy was first being perceived as a salient public issue.

From the perspective of fifty years, much of their message bears an ironic ring in relation to subsequent practice. The expansion in large-scale domestic surveillance over this period—including the rise and growth of agencies like the Department of Homeland Security, the FBI, the National Security Agency, the proliferating "Fusion Centers," and others—appears to have sharply increased the net extent of *secret* recordkeeping on Americans. These shadowy operations are hardly dedicated to disclosing what files they keep or on whom they keep them. The FBI, for example (as chapter 3 of this book illustrates), has gone to special lengths in recent years to exempt itself from Privacy Act requirements to disclose to members of the general public whether they are listed in its files.

Both key government agencies and private-sector suppliers of personal data would recoil against any serious consideration of the third FIP above—the one proscribing use of data collected for one purpose to serve other, different purposes, without the subject's consent. To the contrary, such triangulation of insight into people's lives, based on comparisons of data on the same person drawn from multiple sources and contexts, is basic to what many of these organizations offer to their customers or clients.

Other themes from the five FIPs have seen wider adoption in practice. The recommendations for openness and accountability in response to public inquiries about treatment of one's own data have become routine practice both in the United States and abroad. Of course, much of the reason for such privacy-friendly compliance in this country stems from the incorporation of the Privacy Act of 1974 as the law of the land. Virtually all civilian agencies of the federal government now have well-established routines by which ordinary individuals can request their own records, challenge their contents, or contest decisions based on them. Indeed, when governmental or private organizations find themselves called upon to "add privacy" to their routines for dealing with personal data held in their files, their most feasible gambit often seems to be to open up these processes for participation and verification by the subjects of the data. As chapter 7 of this book notes, recent state legislation in California seems likely to spark a trend toward extending similar disclosure requirements to private-sector businesses in other states.

It would be comforting to take these developments as evidence that the dangers of personal-data systems gone wrong have met their match in these reforms. But that is not my view. Americans and other countries with extensive privacy codes can certainly boast of having more privacy law and regulation on the books now than a half-century ago. But few informed observers would be rash enough to hold that we have more privacy. Over the past five decades, Americans have seen the proliferation of data brokerages, devoted to buying and selling personal data for a vast variety of purposes. We have seen government agencies master the art of monitoring our telecommunications *metadata,* the equivalent of our social DNA—in search of any signs of suspect sympathies. We have seen the country's internet giants develop capabilities of pinpointing within us tastes and susceptibilities we were not even aware we had. And we have seen countless organizations—large and small, governmental and private—perfect and apply means for identifying us and noting our whereabouts and movements via technologies ranging from the "fingerprints" left by our computing equipment (see this book's introduction); to our voices, irises, or gaits (see chapter 1); to the "pings" emitted by our cellular phones (see the introduction). Whatever the virtues of today's privacy protection codes, their supporters have lacked the political muscle to forestall the massive losses of privacy to these and similar innovations.

The procedural, due-process rules so prominently featured in global privacy and data protection codes have every virtue but that of helping decide whether any particular personal-data system deserves to exist in the first place. Fair Information Practices and kindred precepts presuppose that tracking people's lives is a valid undertaking—if only carried out with openness, according to acknowledged rules, and with opportunities for the subject to access and, if necessary, dispute the record. They do not tell us what constitutes valid purposes for creating such systems in the first place, how far they should reach into citizens' lives, or what means of accumulating such crucial information should be held valid. One wonders, for example, whether application of these standards would yield approval of the imposition of forced tests of women's virginity, as practiced under some Islamist regimes. Would this kind of rule-bound, openly acknowledged set of standards satisfy the requirements of privacy in such investigations? Would affording the targets of routine tracking practices access to the records of these procedures, and the right to challenge conclusions based on them, suffice to establish their acceptability? Or should the whole notion be rejected as an exercise in brutality and overweening power, no matter how it is carried out?

My Argument

This book seeks to put forth clear alternatives to prevailing thinking on the defense of privacy values. I do not approach today's burgeoning systems of human tracking and influence as a sort of analytical travel agent—asking readers whether they wish to go by car, plane, or ship. Instead, I want to inquire whether we have set off to the right destination in the first place. If I am right, we have only just begun on what promises to be a long and historic journey, in our reflexive quest for intensified monitoring and reshaping of human behavior. The goal of getting people to behave "better" is an ever-present fact of social life, not to be disparaged categorically. But so are needs to keep power and authority dispersed, accountable, and limited. Our deep-going cultural obsession with compiling more and more personal information—and using such information more artfully to shape the behavior of those so targeted—draws us into a path of profound transformation. As we move farther along that path, we catch sight of an ultimate destination consisting of life with no private spaces whatsoever.

Only self-conscious, critical public reflection can enable us to step back from the prevailing cultural mystique of knowing more and more about people, in order to influence them more efficiently. Hardly anyone wants to live in a world where any personal information, released or entrusted to any party for any purpose, becomes subject to appropriation and use by any interested institution for any other purposes. That proto-totalitarian outcome is exactly what taking privacy seriously seeks to avoid.

Thus, it is no part of my intent for this book to divine the directions in which "technology" is supposedly leading us—perhaps so that no one else gets there before we do. Nor do I speculate on what demands that mythical force may next make of us. Instead, I seek to ask what the most desirable order of public life would be, if we could shake off the inherited limitations of material inequality, power differences, and acquiescence to the supposedly inexorable demands of an Information Society. I seek to focus on information technology as a potential source of enrichment and affordance to democratic pluralism, rather than a series of inflexible constraints. If we were to start from the beginning to create the best possible social order, rather than simply reacting to privacy outrages, what role would we choose for the information technologies at our disposal?

To create conditions for a journey to a destination more desirable than our current one, this work proposes eleven reforms. These are not technological fixes or new strategies of information management, so much as changes in basic social relationships. They set down a strong diet of new individual rights over the treatment of personal data and a demanding new set of responsibilities for institutional keepers of such data. One of the former is a right to "resign" from personal-data systems where inclusion is not legally mandated. Another is a *property right* over one's own personal information, such that its commercial use—in targeting unsolicited marketing appeals, for example—without one's permission would be tantamount to theft. A key new obligation of data-holding organizations would be an absolute stricture against the release of personal data from any system, without legal requirement to do so or permission from the subject.

None of these reforms (listed seriatim in appendix 1) should be conceptually forbidding to American consumers or voters. They have to do with who "owns" personal information; what "downstream rights" people should have over data about themselves, even once such data have become public; and what keepers of personal-data systems must be required to disclose about their activities. But most of the reforms, though not analytically complex, are *excruciating* in terms of the political dramas that any serious attempt to enact them would trigger. Some would threaten major American industries—the data brokerage industry, for example. Others, like the right to resign from data systems, would create useful legal tools enabling individual citizens and consumers to demand better treatment from users of their data. The net effect of applying all these reforms, I argue, would be *synergistic,* with each supporting the others, directly and indirectly. Collectively, they would strengthen the position of all those confronting the entrenched organizations that now control our data.

Adoption of the proposed reforms will certainly come at a cost. Some services—notably consumer credit—are apt to become at least somewhat more expensive, if credit reporting renounces some of its more relentless intakes of personal data. Similar losses could be expected in the collection of some taxes if the authorities reduce the intensity of their surveillance over taxpayers' lives. Some terrorists and their supporters could escape attention, if investigators exercised more restraint over methods used to monitor their movements and communications. All these possibilities must play a role in any extended public discussion of the ultimate privacy question: What kind of informational world do we want to create for ourselves and our descendants?

Some readers will dismiss these suggestions as too remote from any possibility of immediate realization. They will insist that personal information and the prerogatives of manipulating it are such valuable assets in today's world that any program for restoring control to the individuals concerned is simply naive. I disagree. These reforms may not be on the agenda for tomorrow or even next year. But considering the sweep of social change in other realms, and the documented discontent of Americans with prevailing uses of information about themselves, a quantum jump in public opinion on these matters may not be such a fantastical possibility. Even casual observers must have been impressed by the profundity of changes in public attitudes toward issues like same-sex marriage and climate change, within just a few decades. Now California, seemingly always in advance of other states, has recognized a right of consumers to block the sale of "their" information—a step toward the point where it might really be "theirs."

Perhaps, then, the most serious obstacle to stronger, better-organized public support for privacy is a failure to take that possibility seriously.

Notes

1. Brendan Keefe and Lindsey Basye, "Qualified Immunity: Police Off the Hook for No-Knock Raid on Wrong House," 11 *Alive*, Sept. 11, 2020.

2. Kevin Sack, "Door-Busting Raids Leave Trail of Blood," *New York Times*, Mar. 18, 2017.

3. *Records, Computers, and the Rights of Citizens: Report of the Secretary's Advisory Committee on Automated Personal Data Systems* (Washington, DC: U.S. Department of Health, Education, and Welfare, 1973), 41.

Acknowledgments

Books are like living things: they draw their character both from their DNA and from the environments in which they develop. If the author's original concept plays the role of genetic inheritance, then the influences of outside thinkers and institutions during the work's gestation make the environmental contribution. At the moment when this book is launched, I take great pleasure in recording my gratitude to the institutions and individuals who have helped nourish the work, so to speak, in utero.

First the institutions. An enormous plus to my efforts over the nearly six years of preparation of this work has been what I consider an ideal institutional home for such inquiry—the Center for the Study of Law and Society at the University of California, Berkeley. Founded in 1961 as a kind of halfway house between legal scholarship and social science, this distinguished institution continues to foster studies that shine bright light on the actual workings of law in its larger social context, and the inputs of these social processes on the law. Here I extend my particular thanks to members of the core faculty at the Center, including Jonathan Simon, Malcolm Feeley, Catherine Albiston, Calvin Morrill, and Sarah Song. I have a special debt of thanks to Pamela Erickson, executive director of the Center, for her exercise of advanced bureaucratic skills on my behalf. My greatest gratitude here is to Lauren Edelman, our late friend and colleague. Tragically, this acknowledgment is posthumous. The entire Center community shares my grief regarding her unexpected death in the early weeks of 2023.

Further crucial institutions include the UC Berkeley School of Law in general, and within it, the Berkeley Center for Law and Technology (BCLT). These two bodies graciously opened many of their programs and events to me—a further source of enrichment to this book.

In terms of collegiality, the Berkeley campus is extremely rich in thinkers concerned with personal information, organizations, information technology, and privacy. Among those who have provided crucial stimulation and support for this work are Chris Hoofnagle; Jim Dempsey (former director of the BCLT); Deirdre Mulligan, professor in the School of Information; and Paul Schwartz of the Law School.

And beyond brick-and-mortar institutions, there is the Invisible College composed of what I call Privacy Watchers all over the world. The rising consequences of the mass treatment of personal information have given rise to conflict and soul-searching worldwide, along with much educated speculation over varying directions for possible responses in law and policy. This ferment has brought those of us concerned about these possibilities together in extended discussion and debate. Given the global distances involved, we may not meet in person as often as the members of the original invisible colleges, but we certainly draw on one another's thinking. It's been my good fortune to profit from the collegiality of Lee Bygrave (Norway), Ann Cavoukian (Canada), Carole Bailey French (USA), Serge Gauthronet (France), Robert Gellman (USA), Graham Greenleaf (Australia), Kevin Haggerty (Canada), Sarah Igo (USA), Wolfgang Kilian (Germany), Torin Monahan (USA), Michael Musheno (USA), Richard Re (USA), Priscilla Regan (USA), Beate Roessler (The Netherlands), Marc Rotenberg

(USA), Lee Tien (USA), David Vaile (Australia), David Murakami Wood (Canada), David Wright (UK), and many others. My sincere thanks to all of these well-informed and helpful cosmopolitans.

Needless to say—I hope—my gratitude in these matters must never be taken to suggest that any of these friends and colleagues is implicated in any flaws or shortcomings that the reader might find in the following pages. Many of these thoughtful interlocutors have done their diplomatic best to inform me of statements and lines of thinking that they regard as ripe for intellectual euthanasia. It hasn't always worked. The responsibility is my own.

The focus of the last chapter in this salute is perhaps the least widely known outside Berkeley: URAP, the university's Undergraduate Research Apprentice Program, and its director, Stefanie Ebeling. This extremely well-conceived and well-organized program essentially plays the role of broker, bringing undergraduate volunteer researchers and those seeking help together. The role of the apprentices from this program in the creation of this book can hardly be overstated. In this case, the research involved fact-checking statements made in the text, creation of endnotes, and surfing the internet and other electronic and conventional sources for materials that appear in the pages to follow. A number of URAP students, after signing on early in their undergrad careers, remained through graduation. Other veteran apprentices, no longer able to include URAP in their official university schedules, continued to donate their indispensable labor for months and even years after they were no longer officially enrolled. Some have weighed in with crucial facts and connections right through the final preparation of the manuscript.

Some of the longest-serving of these enthusiastic, talented, and tireless people are Pooja Bale, Bani Bedi, Randy Cantz, Leslie Chang, Hank Cheng, Alain Jiang, Zoe Kielly, Bailey McHale, Rachna Mandalam, Julia Ornell, Celia Ruelas Ramirez, Angelica Vohland, and Yao Xiao. Most tasks were shared, but appendix 2 is based largely on the work of Rachna Mandalam and Angelica Vohland; table 1 (in chapter 3) is largely based on the work of Randy Cantz and Pooja Bale; and table 2 in (chapter 6) largely represents the work of Angelica Vohland, Bani Bedi, and Hank Cheng. Without them, this book would have been impossible.

And thanks, posthumously, to Ronald Dworkin, for my borrowing from the title of his distinguished work *Taking Rights Seriously*. I am no Dworkin, but it always helps to set the mark high.

And finally, my heartfelt thanks to the editorial team consisting of Naomi Schneider and Aline Dolinh of UC Press, and to private consultant Dimitrije Stankovic, for their expertise in moving the project from a set of raw possibilities to the work you see here.

James Rule
Kensington, California
May 2023

Introduction
Careering Down a Road Hardly Anyone Wants to Take

It's hard to find anyone willing to predict a bright future for privacy in America. Some simply claim indifference to the fate of what they decry as an anachronistic value. But many are quietly alarmed to see us speeding headlong down a road that few really want to take, but which no one seems able to quit. At the end of this road lies a world where all our recorded information flows seamlessly from one interested party to the next—regardless of our wishes. Few of us, I believe, welcome this trajectory.

Propelling us in this direction are the ingenious and precisely targeted efforts of *organizations*—resourceful government agencies and corporate bureaucracies whose place in the sun depends on what they can do with our data. These parties make it their specialty to master the details of our pasts in order to orchestrate our futures. Sometimes we experience their attentions as uncomfortable, unilateral impositions—as when tax authorities demand documentation for obscure details of our lives, with threat of penalties if we balk. Elsewhere, we uneasily find ourselves complicit in undoing our own privacy, while downloading smartphone apps, installing communications software, or logging on to social media. As we click through opaquely written privacy notices, terms of service, and user agreements, we sense—often correctly—that we are yielding much more than we know. Strictly speaking, we may have the option to step back—to renounce the connection or the service at hand. But since the activities in question are increasingly defined as part of normal life, few choose to turn away.

Even those of us determined to guard access to our personal data find that resistance doesn't get us very far. We live in a world organized unobtrusively to extract and record pertinent information about our whereabouts, our states of mind, our consumer choices, and our political attitudes without our consent—and often without our knowledge. The prescriptions we order, the topics and timing of our phone and email communications, the parties we communicate with; details of our domestic and international travels; even our choices of X-rated videos—these are just a few of an enormously long and constantly growing list of crucial personal data that we shed. Any and all of these fragments of our biographies are subject to being sold, traded, or otherwise directed to government and private organizations dedicated to using them to shape their treatment of us. The fact that we cannot censor information about ourselves collected in these ways makes it especially valuable to organizations seeking it. For such unguarded disclosures are apt to afford the most telling insights into what we're apt to do—or think or want—next.

Countless organizations on which we rely to get on with our lives crave to know things about us that we may not care to share. Credit grantors take an interest precisely in those past accounts where our payment patterns have been spotty or worse. Insurance companies prefer to sell coverage to customers who never even *think* about making a claim—let alone those with a history of doing so. The Internal Revenue Service wants to know about windfall income and well-paid special assignments that taxpayers are least likely to volunteer on their returns. Such tensions, and countless others like them, have been with us far longer than computing. What information technologies have added is the possibility of recording—and sharing, analyzing, cross-checking, slicing, and dicing—the fine details of our daily lives in ways that reveal what would otherwise remain obscure. What's more, realizing the rich payoffs that ensue from cross-referencing self-generated information about us to that derived from outside sources, organizations develop *algorithms* that point to connections any one investigator is unlikely to make on her own.

As law professor Frank Pasquale points out, one never knows when some imaginative combination of personal data may start ringing alarm bells where they're least expected. He offers as an example "the plight of Walter and Paula Shelton, a Louisiana couple who sought health insurance. Humana, a large insurer based in Kentucky, refused to insure them based on Paula's prescription history—occasional use of an antidepressant as a sleep aid and a blood pressure medication to relieve swelling in her ankles. The Sheltons couldn't get insurance from other carriers, either. How were they to know that a few prescriptions could render them pariahs? . . . But since then, prescription reporting has become big business: one service claimed reports of 'financial returns of 5:1, 10:1, even 20:1' for its clients."[1] The author notes that the Affordable Care Act may now shield the couple from discrimination like that described here, through its protections for patients with preexisting conditions. Chapter 2 of this book explains more fully how data on their prescriptions captured from this couple's pharmacists may have tipped insurance companies to avoid them.

The drip-drip-drip accumulation of personal information resulting from such connections may pass without notice in a world where computer monitoring of our actions has become a daily, if not hourly, experience. Still less obvious is the growing power of those who control the systems—from social media to government watch lists to marketing databases—as they broaden in their coverage, and fuse with one another in ever-new combinations and symbioses.

Such emerging forces in American life are not simply manifestations of technology. These phenomena are *political* both in their inspiration and in their consequences. Changes in who-can-know-what-about-whom inevitably bring transformations in power and authority. Taxation, for example, is always dependent on availability of information on the people and things to be taxed. And it avails the authorities little to declare a levy on consumption of internet pornography, for another example, unless means exist to identify and track such content. Such surveillance, though, is by no means beyond the capabilities of current technologies. Is America ready to accept a "sin tax" on porn—to go with extra charges levied against tobacco and alcoholic drinks?

And have Americans fully come to terms with the possibilities of using—or abusing—such surveillance capacities to distort basic processes of democracy? Nearly everyone remembers the revelations of attempted misuse of personal data in the 2016 presidential election: Cambridge Analytica, a shadowy British company, obtained data on millions of Facebook members, with the intent of manipulating their votes in the presidential election.

Attempts to distort expressions of opinion between the public and government by no means ceased with those events. In an effort to inform its expected ruling on net neutrality, the Federal Communications Commission opened its website to public comments. According to an account from April 2018, it received more than twenty-two *million* comments. But close investigation of such efforts to canvass public sentiment on major policy issues has documented stunning efforts to manipulate the results. The *Wall Street Journal*, for example, dug into the backstory behind a Labor Department forum involving public views for and against changes in the "fiduciary rule" requiring retirement advisors to act in the best interests of their clients. A difficult principle to oppose, one might have thought. But the *WSJ* analysis of the "public comments" on the proposed changes found that some 40 percent of them were fake—posted under the names of real people, but not by those who allegedly authored them.

Many of these false public voices appeared to be impersonations of those whose identities had been compromised by at least one data breach.[2]

. . .

Many of these possibilities for deception and manipulation remain largely unknown, even to those with long experience in these matters. *Washington Post* writer Geoffrey Fowler has published an in-depth inquiry into the "fingerprinting" of computers—techniques by which organizations can determine the identities of parties they are interacting with—even when the latter take all available measures to cover their electronic tracks. Their computers can force yours, Fowler reports, to give up information on matters like the resolution of your screen, your operating system, and the fonts you're using—technical details, perhaps, but distinctive enough in the *ensemble* to reveal your identity. "At least a third of the 500 sites Americans visit most often use hidden code to run an identity check on your computer at home," he concludes.[3] Thus, companies devoted to providing pornography over the internet may know your preferences in this form of entertainment before you even log on—inductively, one might say, simply by generalizing your taste in this medium as manifested across many visits.

Yet such subtle invasions of privacy pale in comparison to what today's personal-data systems could do, if mobilized under an overtly repressive regime. They could track the whereabouts of political dissenters, for example, or disable their bank accounts, or pressure their friends. A regime with totalitarian intent, abetted by present-day technologies and management techniques for manipulating personal data, would create a "perfect storm" for democratic institutions. Without strong legal and policy checks against such uses, we could find ourselves approaching authoritarian rule through the back door—developing an arsenal for intrusive scrutiny over once-private areas of life that would go unnoticed until used for the most destructive purposes.

. . .

American law has never developed a unified philosophy of privacy. This is scarcely to say that we have no significant protections for personal information, but rather that restrictions on what can be learned about and done with personal data derive from disparate sources—some from the Constitution, some from tort law, some from criminal law. Fourth Amendment constitutional protections against state powers to obtain certain personal records block some forms of intrusion into people's homes. But the fact that so much personal data is stored outside Americans' homes—in banks, credit card account records, and the like—greatly reduces the overall significance of the Fourth Amendment as a source of privacy protection muscle.

The closest this country comes to European-style privacy legislation grounded in a clearly defined "right to privacy" and applying a single set of principles to an open-ended variety of cases is the Privacy Act of 1974. This key statute governs treatment of personal data held by U.S. government agencies. In the underlying principles it invokes, it hearkens directly to the influential publication *Records, Computers, and the Rights of Citizens* (1973)—the source of the much-noted Fair Information Practices that have since influenced so much legislation and policy throughout the world (see this book's preface).

Here and elsewhere, the rise of large-scale personal-data systems designed to guide government or corporate action toward "private" citizens has fostered a new *genre* of jurisprudence. Since the 1970s, legislators and policymakers both in this country and abroad have sought to constrain the use of personal information in the face of these new modes of action. These efforts have resulted in the proliferation of national privacy codes—now more widely termed "personal data protection" laws to an estimated more than two hundred countries outside the United States. First in Northern and Western Europe, then increasingly throughout the globe, these bodies of law have come to shape practice both in government and in the private sector.

But America's privacy codes have afforded little defense against the salient privacy setbacks of the last generation—the rise of industries devoted to compiling and retailing Americans' personal information, for example; or the intrusive monitoring of telecommunications metadata by shadowy government agencies; or the rise of social media, with their weak privacy guarantees and their determination to market insights into personal attitudes and susceptibilities that we may not be aware of ourselves.

We can do much better. America's efforts to protect privacy in the face of encroaching demands for personal information have not matched the enormity of the forces that they are supposed to address. Indeed, American legislation and policy have often approached these issues on the basis of potentially disastrous assumptions. One is that any existing system of accounting and tracking of individuals' lives—government or private—must correspond to some authentic, shared social *need*. In this view, the reformers' task becomes one of *adding privacy* to existing practices, without threatening the key performances of the organizations involved. The resulting measures may indeed succeed in granting the subjects of the recordkeeping some significant relief. But they leave the most historic and consequential questions unanswered. Above all, how much and what forms of such human tracking deserve a place in a liberal and democratic social order in the first place? And what can we do to reduce the extreme asymmetries in options for action between data-holding organizations and grassroots citizens and consumers? In other words, what steps can we take to make Americans not just subjects of institutional recordkeeping, but instead active and effective agents of their own privacy interests?

This book takes such questions as its point of departure. It seeks to break with a tradition of American privacy protection that, for all its virtues, has proved excessively *reactive* to the ever-emergent stream of privacy outrages. Instead of asking how we might somehow *add privacy constraints* to existing personal-data systems, I want to pose what strike me as prior questions. In the best of possible worlds, what uses of recorded personal information would serve us most efficiently? What legal and policy principles would someone who takes privacy seriously want to uphold—and what practices should simply be abandoned?

. . .

How can we define *privacy* for the purposes of this book? This usefully vague term refers to a mixed array of human interests that are clearly akin to one another, yet distinct. One of these is a felt need for *autonomy* in decision-making and action—the insistence on respect for realms where one makes up one's own mind and charts one's own course of action. People often assert this form of privacy interest in matters of birth control and abortion. Here as elsewhere, an interest in keeping certain knowledge to oneself stems from broader desires not to have to account to anyone else for crucial decisions. So, the ruling in the famous court case *Griswold v. Connecticut*—that the state could not ban the use of contraceptives by married people—is characterized as a victory for marital privacy, in this sense of *autonomy*.

Another basic human desire often understood as a privacy interest is the wish to avoid embarrassing *exposure*. Hardly anyone likes the prospect of being forced to disrobe when all others present are clothed, or being spied upon when assuming that one is alone. We have similar desires to avoid publication of details of medical treatments that we find awkward, or public review (in court proceedings, for example) of incidents that we consider shameful or embarrassing. The law, in many states, recognizes this privacy interest—as in statutes prohibiting publication of the names of rape victims or dissemination of revenge pornography.

Information privacy, people's interests in institutional compilation and use of data held in file about themselves, has attracted the lion's share of commentary over the past five decades—no surprise, given the skyrocketing growth in personal-data systems and their consequences for the lives of those depicted in them. Nearly everyone today takes it for granted that their "records" follow them through life, with weighty results—even if one is unable to specify fully what those records consist of, or precisely to detail their uses. Information privacy—often shading off into these related forms—is the main focus of this book.

Keeping these distinct meanings of privacy in mind wins us a measure of precision. But I nevertheless believe that our use of the vague but inclusive term *privacy,* with its multiple, overlapping meanings and connotations, has value of its own. We need to think of privacy as what philosopher W. B. Gallie called an "essentially contested concept"—an idea destined to be argued over and reassessed as long as issues of autonomy, exposure, and treatment of recorded information matter in human affairs.[4] As with "liberty," "justice," "good citizenship," or "democracy," understandings of privacy inevitably have a normative element, in that most people consider a certain measure of it, at least, a good thing. But if privacy in some sense is indeed desirable, then what policies and practices most effectively promote it—or promote the "right" kinds of privacy, or privacy in the right contexts? Most of us would not opt for granting unlimited privacy to former felons regarding their criminal histories, for example. So, how do we distinguish appropriate and indispensable forms of privacy protection from those that would wreak undeserved hardship? These are classic questions raised by essentially contested concepts.

We can't expect definitive settlement of such questions. Perhaps this is because every age and every intellectual constituency brings different concerns to the debate, so that what counts as a satisfactory understanding for one thinker at one moment will miss the mark for others. When we speak of privacy, we point to an inchoate complex of values and interests that is constantly undergoing redefinition and reinterpretation. Perhaps this is unsatisfying. But we can't get along without essentially contested concepts—or at least, we shouldn't want to—because struggling to define and apply them anew is essential to our efforts to make the best of the world we live in.

. . .

What do I mean, then, by "taking privacy seriously"? I mean, above all, placing privacy on a par with state security, institutional efficiency, and other social goods with which it inevitably competes. I mean being willing to renounce some—by no means all—of the securities and comforts afforded by today's personal-information systems in the interests of privacy. I mean being prepared to run the risk of dialing down some—not all—of the intrusive surveillance measures that today fly under the flag of national security. And I mean being candid about the costs that must be paid for refusing to exploit every bit of personal information for institutional ends—for example, in giving some poor credit risks a "second chance" to borrow and repay, or allowing a former convict who has broken with his or her past to live quietly and without discovery.

This needs some unpacking. I hardly imagine that every reader of this book joins me in embracing all the privacy-friendly priorities cited above. Some will proceed from quite different assessments—for example, the notion that the world is so dangerous and disorderly that intensified supervision and closer discipline of citizens' lives is the only reasonable policy. Others might hold that privacy values are ultimately a matter of indifference to most people, most of the time, and hence do not warrant the significant costs needed to uphold them. Those readers will approach the following pages as a road map for a trip they do not wish to take.

What can I say to persuade those readers to share my view? I take it as axiomatic that differences of political vision like these are only partly susceptible to resolution through appeals to evidence—if by "evidence" one means empirically observable data. Often such conflicts in worldview more closely resemble differences of *taste* in art, recreational habits, or even food than disputes over matters of empirical fact. We can and must be prepared to argue for the *workability* of any values or tastes that we choose—that is, for their ability to direct and sustain a social world we would be prepared to live in. But we should never deceive ourselves that the role of such tastes can somehow be eliminated. Some citizens—not too many, I suspect, but some—simply feel little attachment to privacy values, at least as compared to competing values. The latter include strict adherence to social rules and rigorous defense of efficiency—from punishing welfare cheaters and tax evaders to making the trains run on time. But they, like we, must offer their fellow citizens some account of how their (or our) favored values could work in practice to animate a livable and attractive world.

Still, I believe that I am far from alone in embracing the key values guiding this work. Few of us are really comfortable contemplating a world where, like it or not, everyone generates more and more recorded personal information simply in the course of everyday living—and where any and all such information automatically becomes available to interested decision-makers elsewhere. Against that view, those of us willing to take measured risks on behalf of privacy need some sort of program of public action. In their necessarily *reactive* role in decrying obvious privacy outrages, American privacy advocates have often missed opportunities to project positive, comprehensive statements of what they are *for*. What sort of world would we want to create, if we had full liberty to fashion institutions and practices with privacy as a central value?

The privacy values at issue are of two kinds. First are *strictly individual goods* implicated in the flow of personal information. These range from those at stake in dissemination of painful or embarrassing facts—as in disclosures from sensitive medical records, or in revenge pornography—to revelation of information that is strategically disadvantageous, as in release of genetic information that blocks access to employment. What these interests or values have in common is that they may be satisfied for one person, yet unfulfilled for the next—just as one may be well fed, for example, while one's neighbors starve.

But a second kind of privacy value also requires our advocacy—those values that can only be experienced jointly, as part of some larger group. One cannot enjoy the satisfactions of nationalism, for example, without a nation to belong to. Or—as is often said—one cannot fully enjoy the virtues of freedom of expression, unless others do the same. A major payoff of such freedom, after all, should be the benefits derived from full exchange of ideas among one's fellow citizens—something impossible if any group feels constrained against speaking out. Privacy is like that. The chilling effects generated by a conviction that large institutions know everything about us can be far-reaching, whether that belief is well founded or not. Accordingly, we all have an interest in *each other's* privacy—in a social order where no one need fear oppressive snooping.

There is nothing contradictory about pursuing both these goals in a single program. Safeguards that we might desire for ourselves are equally valid when sought on behalf of our fellow citizens. Under prevailing American law, personal data are largely governed by what might be called "finders-keepers" principles, at least with regard to commercial use. As the following chapters demonstrate, information ranging from the details of our internet searches to the prescriptions we order from the pharmacy are subject to opportunistic capture, reuse, and resale by organizations we trust to handle them, or those that pay such organizations for access to the data—hence Americans' widespread, and largely justified, perception that control over one's "own" data, once released, is permanently lost. The reforms proposed here seek to reverse this state of affairs. They even go so far as to propose (in chapter 6), a universal property right over commercialization of information on oneself, so that no such use would be possible without explicit consent from the subject.

Any attempt to enact such reform, needless to say, will trigger tectonic collisions of interest with the parties now controlling commercialization of our personal information. Simply injecting this possibility into public discussion would send an unmistakable message that today's regime for the use of personal data is not the unique or inevitable consequence of prevailing technologies. It is exactly the sort of easily understood, politically potent possibility—like Medicare for all, or paid family leave, or a direct tax on carbon emissions—that could fuel a movement endowed with a positive social vision of a drastically more private world. The mere fact that such alternate realities can be acknowledged and discussed refutes one of the most dangerous anti-privacy themes of current culture—the notion that today's privacy-challenged practice represents the only possible result of our reliance on computing technologies.

This book proposes eleven potent and easily grasped privacy-friendly principles, and seeks to weave them into a comprehensive vision. The scarcity of such visions, I hold, has undermined possibilities for a strong, grassroots movement on behalf of privacy in American life. Americans have heard so many times that they have "chosen" a more convenient, diverting, comfortable informational world over a more private one that many have begun to believe it.[5] But that notion does not withstand close examination. Who among us has not grown jaded in the attempt to make sense of sweeping terms like those in the following, outrageous agreement proposed by Google and signed off or clicked through despite—or perhaps because of—the asymmetry of the situation: "You retain copyright and any other rights you already hold in Content which you submit, post or display on or through, the Services. By submitting, posting or displaying the content you give Google a perpetual, irrevocable, worldwide, royalty-free, and non-exclusive license to reproduce, adapt, modify, translate, publish, publicly perform, publicly display and distribute any Content which you submit, post or display on or through, the Services. This license is for the sole purpose of enabling Google to display, distribute and promote the Services and may be revoked for certain Services as defined in the Additional Terms of those Services."[6] What pass as "choices" of this kind are not authentic acts of choice. I would describe them more as chewing gum for the mind.

As a substitute for such manipulated "choices," I invite us all to step back and consider the full range of differences between the unprivate world that we inhabit and a world where privacy values bulk at least as large as those of expedient service to institutional interests. Without such a vision, any social movement faces great difficulty in mobilizing the energies of would-be followers—even in the face of manifest dissatisfaction with the status quo. "You can't beat something with nothing," the old political adage has it. In this case, you can't beat a world that seems to be flourishing at the expense of privacy, without an appealing vision of an equally livable world that incorporates strong defense of that value.

True, even those I consider my natural allies in these matters will not subscribe to all the proposals put forward here. I have deliberately cast the arguments that follow in polemical form, and many of the reforms advocated here will strike some readers as utopian. This is intended—though *utopian* has unfortunately become a term of dispraise among many thinkers on policy, as if synonymous with *unrealistic* or *unsophisticated in the realities of implementation*. In fact, a key aim of this book is to uphold utopian thinking in what I take to be its classic sense of exploring the best possibilities of human relations, *absent* the hindrances and limitations imposed by power differences, constraints born of ignorance, and other accidents of history. This form of thought, I hold, offers a bracing and provocative opening to discussion of real-world public affairs—so long as it does not blind us to the distance between where we are and where we need to be, or to the power of the forces militating against closing the gap between the two.

I hope this latter objection can never be posed against the arguments that follow. Throughout the book, I seek to emphasize the intensity of the political winds blowing against the reforms proposed here and, indeed, to call attention to some that the reader might miss. I seek to concentrate minds on the fact that the privacy-devouring character of current institutions and practices is no more than that, and not inherent in the nature of today's technologies or, still less, in human nature itself. Such thinking, as students of social movements have long held, provides an indispensable catalyst for organizing efforts. Identifying concrete changes that hold the promise of a much better world—even where those steps appear enormously demanding—can thus represent the beginnings of meaningful change.

Nevertheless, some readers will gasp at these proposals, if only because many are so sweeping. Some of the reforms proposed here would be relatively simple to institute, as matters of data-management policy—for example, establishment of a universal "right to resign" from any personal-decision system not required by law (as advocated in chapter 4). But other recommendations would be both complex, time-consuming problems for implementation *and* minefields of political opposition—like the creation of two master computer portals to provide universal access to the workings of most major personal-decision systems (as advocated in chapter 3). A number of such reforms, besides being ambitious and costly, would take years to plan and carry out.

Note, though, that I propose all the reforms advocated here not as short-term expedients, but as long-term goals and talking points for a privacy movement oriented toward fundamental change. Underlying these proposals, always, is the notion that no social movement can hope for success without some overarching vision of a better world as the target of its strivings. By the same token, many of these reforms raise highly contentious issues in the law—for example, the status of personal data protection in the face of First Amendment guarantees of freedom of expression. I seek to show that satisfactory compromises are also achievable here, despite weighty political obstacles. Thus, I ask readers to weigh the value of these proposed reforms in their own terms—for the contributions that they might, at length, make to building a much more private world—rather, say, than by the likelihood of their adoption in the next session of Congress.

The real opposition here will come not from other privacy advocates, but rather from those for whom these questions are settled in advance. Such beliefs take a number of forms. One of them might be dubbed First Amendment fundamentalism. This is the conviction, buttressed by a new wave of conservative jurisprudence and aggressive neoliberalism, that transmission of virtually any information about human beings constitutes *speech*—and thus deserves the broadest of First Amendment protections. These doctrines have given us the notorious Supreme Court ruling in *Citizens United v. Federal Election Commission*, among others. The *Citizens United* ruling struck down limitations on political campaign expenditures, on the grounds that these measures restricted the expression of political ideas in elections. If voting is indeed a form of speech, and if money indeed helps votes to flow freely, then it follows that limitations to flows of cash to promote voting on behalf of particular candidates or principles infringe on First Amendment rights.

Similar extensions of First Amendment thinking have led some to entertain the possibility that *virtually any* privacy legislation could be seen as a curtailment of free expression. The stakes, these thinkers hold, are of the highest sort: "If the legal system accepts the propriety of laws mandating 'fair information practices,' people may become more sympathetic to legal mandates of, for instance, fair news reporting practices or fair political debate practices."[7] For thinkers of this frame of mind, such government-enforced obligations of "fairness" in media treatment of political debate or other news would be tantamount to demands for thought control.

Other rationales to forestall taking privacy seriously as the basis for grassroots action invoke the *special expertise* supposedly required of anyone who would challenge high-tech activities. Only those with the most advanced understandings of the latest information technologies, these arguments go, can appreciate how deeply dependent we all are on the "free flow" of personal data. Curtailing that easy access to the details of people's lives would, accordingly, make about as much sense as outlawing the burning of coal to produce steam just as the Industrial Revolution was getting under way.

My position is the opposite. I believe that the most important choices facing us regarding the future of personal information are essentially political—that is, choices of *what kind of social world* we most wish to inhabit. Does it make sense to trust the forces now controlling the appropriation and uses of "our" information—now, and into the indefinite future? Or is it our responsibility to fashion reasoned limits to how readily information about our lives flows, to whom, and for what purposes? Such questions are bound to be agonizing, given the entrenched position of the interests now dominating the use of personal information. What is at stake in the privacy wars is no more complicated than in matters of climate change, or gender equality, or the dangers posed by global criminal enterprises. These questions are challenging, but certainly accessible enough to thoughtful members of the general public to afford meaningful debate.

Perhaps what strikes us as *complexity* here is simply the astonishing or unintuitive ways in which personal information can currently be obtained and exploited. Consider *geofencing*—a novel and chilling technique for spotting, tracking, and influencing supposedly anonymous people in public places by capturing the "pings" emitted by their cell phones. The uses for this information are almost as stunning as the means for its collection. Without alerting the members of a crowd—assembled, say, to hear a speaker of a specific political bent—the geofencing party "harvests" the phone numbers and the location histories of all those present. Then it can go on to record whom these people socialize with, where, and on what occasions.

From such data, experienced pollsters can readily guess whom each member of the crowd will be supporting in upcoming elections. Those diagnosed as "friendly" in this respect can be targeted for telephone calls offering them reminders on election day, and rides to the polls. Needless to say, those whose geofencing analyses show uncongenial political views cannot expect any such overtures. "People might get freaked out about this technology," said Democratic political strategist Kimberly Taylor, who reported relying on some closely related technologies to track both potential political allies and antagonists. "But we're trying to use it for good."[8]

No doubt, all those who avail themselves of such privacy-eroding capabilities believe they are using them "for good."

One hears much, these days, of the degradation of American public debate on great issues of the day. We have fewer and fewer independent sources of political information to rely on, it's said, and more incentives to derive all the information we do absorb from sources whose political orientation resembles our own. In geofencing we have a set of technologies and practices that intensifies these trends—encouraging political strategists to identify potential voters of given mind-sets, and to direct their appeals exclusively to them. No wonder there are so many complaints that American political talk resembles an echo chamber, or that political communication and action have grown more polarized.

But there is worse. The forces driving these changes work in tandem with headlong destruction of Americans' privacy—as expensive, high-tech appropriation and manipulation of personal data comes to replace direct communication across electorates as wholes. In classic democratic theory, elections operate as impartial markets in political ideas and directions. Thus, voting resembles well-informed shopping. Discriminating voters reward candidates and parties that promise the best support for voters' well-considered ends. But with geofencing and related technologies, matters are reversed. Well-funded strategists vie with one another to select the voters who will be encouraged and enabled to speak for entire electorates—for example, by receiving special encouragement, and special access to the polls in the form of sponsored rides. Instead of broadcasting appeals to all members of the polity—that is, to assured supporters and critics alike—today's high-tech strategists seek to appeal selectively, by sending what amount to back-channel messages to their own special constituencies, as identified by geofencing and the like.

Ingenious manipulations of personal information like geofencing are fascinating in themselves, if not particularly inspiring. But the broad message of this book has to do with social and political relationships—with attention to the role of information technologies in shaping such relationships. What possibilities for collective action arise from new ways of identifying sources of political support like geofencing? How do such developments work to discourage debate across all members of the polity? Or, to put matters more positively, what is the best role that technologies could play in enriching public life in our technology-dependent world? What steps could we take to ensure that further innovations in this realm support and amplify key social values?

And why is it that innovation so often plays quite the opposite role? Why do we continue to fall back on ideas that have demonstrably failed to address the full seriousness of the forces working against us? I am thinking here of ideas like "notice and choice." This doctrine holds that consent for initiating complex relationships between holders and subjects of personal information can be established by a quick check of a box, or a click on a keyboard—or perhaps simply by not aborting an ongoing interaction. Often, when examined closely, the resulting "agreements" can lead to grotesquely disadvantageous and burdensome obligations for users that no one other than a specialist could anticipate. I have never encountered a reasoned argument for the substantive validity of such "consent." The situations in which it is given are so hurried, and the resources of users to analyze the terms so limited, as to mock the very idea of reasoned deliberation.

Any effort to address the pervasive asymmetry between the positions of users and providers of sophisticated information services could do no better than to start here.

Human Costs

Discussion thus far has focused mainly on big structures and broad social trends—the rise of institutions for tracking millions of Americans at a time, for example, and the enforcement practices associated with such changes. But most of us experience such tectonic shifts in who-can-know-what-about-whom through our everyday lives—often in the form of new kinds of interaction between institutions and ordinary Americans. To be sure, these new experiences are often positive. Sophisticated means of linking people to their "records" may make it possible to locate urgently needed organ donors, for example; or to help reunite families separated years before; or to coordinate medical care when vast distances separate patients from physicians.

But at the other end of the spectrum of outcomes, new personal-decision systems often leverage long-standing inequalities and disadvantages, leaving vulnerable categories of Americans even harder-pressed than before. For privacy advocates, the ultimate nightmare scenario in these matters must be intensification of pressures on beleaguered groups who have already suffered from discrimination based on race, gender, or sexual orientation. Consider the case of Carmen Arroyo, an apartment-dweller in Connecticut whose son Mikhail had suffered a debilitating accident that left him in a nursing home. Carmen, who had become her son's conservator, applied to her landlord to move Mikhail from a nursing home into her apartment.[9]

But the application was denied—not, ultimately, by the landlord but by CoreLogic Rental Property Solutions, a contractor retained by the landlord to screen prospective tenants. CoreLogic's report specified that Mikhail's background showed a "disqualifying [criminal] record." Carmen thus confronted what appeared to be an impossible situation. Desperately struggling to care for a seriously disabled family member unable to act in his own interests, she was prevented from taking compassionate steps on his behalf. Her adversary was not a human decider, but a data-driven organization that produced its decisions according to secret rules—and that refused, she asserted, to provide her with a copy of the information on which its decision was based. She later learned that her son's only criminal record was a charge of shoplifting, dating to before his disability and subsequently dropped. Nevertheless, Mikhail remained in a nursing home approximately one year longer than necessary. It was as though Kafka's *The Trial* had been rewritten for the twenty-first century, perhaps by the editorial staff of *Wired* magazine.

Fortunately, Connecticut has relatively strong tenancy laws. Carmen Arroyo managed to contact the Connecticut Fair Housing Center and the National Housing Law Project, who filed suit against CoreLogic Rental Property Solutions—for discriminatory use of criminal records as rental criteria.[10] The plaintiffs argued that the company, and all tenant-screening companies, must follow fair-housing requirements when they act on behalf of landlords in making rental decisions on tenants. In early action in the suit, the court found that the plaintiffs had presented sufficient national and state data to suggest that the company's practices had a disparate impact on African Americans and Latinos. It noted that "disparities adverse to African Americans and Latinos and in favor of whites exist at all stages of the criminal justice process: in arrest rates, in jail detention rates, and in prison incarceration rates." The court also denied CoreLogic's motion to dismiss the case, ruling that "CoreLogic was aware that its criminal background screening product could have a disproportionate and arbitrary effect on racial minorities, and that it had taken no steps to modify its product."[11] At the time of this writing, the trial was just going ahead.[12]

Perhaps the most intimidating circumstance facing Carmen Arroyo was the distant and impersonal authority claimed by CoreLogic—the combination of implied claims for scientific objectivity and the lack of any opening to debate the logic of their position. The landlord had conferred her fate to an organization that had both cultivated unknown sources of crucial information on her son (and perhaps on her, as well) and then handed down a judgment as uncompromising as Moses's dissemination of the Ten Commandments. In this respect, her position was not much different from what most Americans face when they apply for credit, insurance, university admissions, dating services, adoption agencies, and even placement in nursing homes. Organizations carrying out these activities devise elaborate point-scoring systems describing the characteristics of the party in question, then make decisions strictly "by the numbers."

The "numbers," in many cases, are the ones that yield the highest possible revenue for the organization making the decision. Thus, in a different investigation, I remember spokespeople for a church-related, officially not-for-profit nursing home describing a point system that they had developed to determine which applicants to admit. A key consideration was how much revenue different mixes of admissions would bring. Virtually all patient fees were paid by government programs, and those fees were inevitably higher for people with what were considered more complicated diseases and disabilities. Thus, the institution went to great lengths to fill its rooms with people with the most complicated illnesses, even if that meant leaving other rooms vacant while less severely ill applicants and their families waited for admission. As far as I could tell, the latter never learned the reason why those not sick enough had trouble gaining admission.

Trends and practices like these surface frequently in this book as drivers of changes that both undermine privacy and subvert people's efforts to fight back against the disadvantages of a privacy-unfriendly world. Organizations often put forward programs for computerization of their paperwork as strides toward efficiency, assuring their customers that their aim is "to serve you better." These claims deserve to be met with all the critical skills we can muster. As in the cases mentioned above, computerization provides fertile ground for formal decision rules that seem to outsiders to leave no "fingerprints" indicating whose interests are served by the changes. Creators of the systems may build in complex scoring algorithms—point systems, in effect, like the one developed by the church-related nursing home mentioned above. Often these formal models generate not just binary "accept or reject" decisions, but also determinations of how much those accepted for business will pay for the services in question. The reliance on quantitative inputs, and the "just the facts" attitudes of those who input data for any particular case, often lend an air of scientific objectivity to the proceedings. But all such systems are created by human beings, and it would be madness not to inquire which humans, and which of their concerns, the algorithms are serving.

For a long time after their original rise, America's big information companies appear to have benefited from a metaphorical blank check of public credulity. Google, Microsoft, and the many smaller organizations that used the services of these giants all seemed to share space beneath the halo of science in American public opinion. Relying on data considered objective, the information industries seemed to claim higher moral ground than, let's say, General Motors or the pharmaceutical industry. "Don't be evil" was a semiofficial motto of Google, and many people both inside and outside the company seem to have taken the statement at face value.

But that sort of moral one-upmanship has now worn thin with many Americans. People have noticed that GAFAM—the name some skeptical Europeans have coined for Google, Amazon, Facebook, Apple, and Microsoft—have their own interests, ones hardly identical to those of their users and customers. Decisions and services provided via complex personal-data systems are no longer beyond the pale of serious criticism in America. Some of that skepticism was evident in the early sparring in the Connecticut case brought by Carmen Arroyo.

Plan of This Book

This book aims to introduce issues like these to readers who may be considering them for the first time—as well as to extend discussion among those long familiar with them. Once understood, the underlying questions should challenge all of us to reflect on some of our deepest and subtlest values. What larger good is served, I want to ask, by supporting seemingly endless intensification of institutional "watching" over once-private areas of life? If we choose—by default or design—to continue the prevailing embrace of more-of-the-same for another fifty years—or a century—will life under those regimes be worth living by the standards of today? If not, might not reliance on the contrarian reforms put forward in the following chapters create some hope of reversing directions?

Chapters 2 through 6 each puts forward a single reform, or a handful of related reforms, aimed at advancing privacy through marked departure from current practice. Chapter 2 calls for imposing a requirement for explicit legal bases for all personal-decision systems, similar to that existing in Europe today. Chapter 3 proposes banning disclosures from such systems, except as required by law or with the consent of the subject. Chapter 4 proposes the creation of two massive web portals, one containing comprehensive information on all personal-decision systems, the other for granting access to one's own data files. Chapter 5 considers a right for anyone to have his or her files deleted from any such system, except where required by law. Chapter 6 proposes creation of a property right over commercial exploitation of data on oneself, such that no organization could use anyone's name or other data for commercial purposes without their explicit and informed consent.

But first, chapter 1 sets the stage for chapters 2 through 6 by charting the rise of surveillance organizations and personal-decision systems from the early twentieth century. It points to parallels among such familiar institutions as credit reporting agencies, criminal record systems, insurance records, and auto and driver registration. It identifies the tracking of individual lives by such institutions as predating the rise of computing—which has nevertheless greatly accelerated its growth. Readers more interested in the historical context of privacy struggles may want to begin with chapter 1. Those whose greater interest lies with the specific proposals for reform may prefer to begin with chapter 2.

Notes

1. Frank Pasquale, *The Black Box Society* (Cambridge, MA: Harvard University Press, 2015), 26–27.

2. Roxanna Ramzipoor, "Data Breaches and Democracy: How Content Abuse May Threaten Freedom of Speech," *Sift* (blog), Apr. 26, 2018, https://blog.sift.com/data-breaches-democracy-content-abuse/.

3. Geoffrey A. Fowler, "Think You're Anonymous Online? A Third of Popular Websites Are 'Fingerprinting' You," *Washington Post*, Oct. 31, 2019.

4. W. B. Gallie, "Essentially Contested Concepts," *Proceedings of the Aristotelian Society* 56 (1956): 167–98.

5. Joseph Turow, Michael Hennessy, and Nora A. Draper, "The Tradeoff Fallacy: How Marketers Are Misrepresenting American Consumers and Opening Them Up to Exploitation," Annenberg School for Communication, University of Pennsylvania, 2015. This penetrating analysis of survey data attributes Americans' acquiescence to the collection of personal data on themselves more to resignation than to any sense of exchange for value in relinquishing such data.

6. Google, "Google Terms of Service," https://policies.google.com/terms/archive/20070416-20120301?gl=US&hl=en.

7. Eugene Volokh, "Freedom of Speech and Information Privacy: The Troubling Implications of a Right to Stop People from Speaking about You," *Stanford Law Review* 52, no. 5 (May 2000): 1049–1124, 1053.

8. Sam Schechner, Emily Glazer, and Patience Haggin, "Political Campaigns Know Where You've Been. They're Tracking Your Phone," *Wall Street Journal*, Oct. 10, 2019.

9. Cohen Milstein, "Judge Advances Fair Housing Case Citing Racial Implications of Criminal Record Screening," Aug. 7, 2020, https://www.cohenmilstein.com/update/judge-advances-fair-housing-case-citing-racial-implications-criminal-record-screening; Cohen Milstein, "Connecticut Fair Housing Center, et al. v. CoreLogic Rental Property Solutions," https://www.cohenmilstein.com/case-study/connecticut-fair-housing-center-et-al-v-corelogic-rental-property-solutions (accessed Dec. 18, 2021); *Connecticut Fair Housing Center v. CoreLogic Rental Property Solutions, LLC*, No. 3:18-CV-705, United States District Court,

D. Connecticut, Judge Vanessa L. Bryant, Jan. 24, 2020, *Leagle,* https://www.leagle.com/decision/infdco20200127a04.

10. *Connecticut Fair Housing Center v. CoreLogic Rental Property Solutions, LLC.*

11. Ibid.

12. On July 20, 2023, U.S. District Judge Vanessa L. Bryant ruled in favor of CoreLogic in this case, while awarding $4,000 damages and attorneys' fees to the plaintiffs.

1 *Don't Blame Technology*

The Enlightenment idea of privacy is breaking apart under the strain of new technologies, social tools, and the emergence of the database state. We cannot hold back the tide.

BILL THOMPSON, British technology writer, "The Death of Privacy and Why We Should Welcome It," remarks made during the Lift Conference, January 18, 2009

The inevitability of transparency is a direct result of an unstoppable arms race in communications tools and data mining operations. . . . We can only adapt to this fact; resistance is futile.

NOVA SPIVAK (CEO and cofounder of Bottlenose, technology futurist, angel investor, serial entrepreneur), "The Post-privacy World," *Wired*, July 2013

We firmly believe that privacy [is] both inconsequential and unimportant to you. If it were not, you probably would not have a Facebook, Twitter, or LinkedIn account: and you certainly wouldn't ever use a search engine like Google. If you're one of those tin-foil-hat-wearing crazies that actually cares about privacy: stop using our services and get a life. We agree with Mark Zuckerberg when he pithily opined "The age of Privacy is Over."

From a "Privacy Policy" statement for the search engine Skipity, 2012

These statements invoke a key myth of our times: the notion that our privacy is being plundered by the impersonal forces of Technology. Like most myths, this one carries a pointed message for conduct: as Nova Spivak inveighs, resistance is futile. Efforts to stanch the flow of our personal information, attempts to assert control over who can access and act upon our most intimate data, are foredoomed. We live in an *information society,* the mythmakers are apt to remind us. Any attempt to interfere with the free flow of this vital input runs the risk of upsetting the very dynamism of our age.

Such thinking conveniently relieves those who embrace it of any responsibility for the unfolding fate of key social values. Yet the fact that epidemiologists have learned how pandemics are born and spread hardly leads us to conclude that they must inevitably set those powers to work. To the contrary, we expect experts to exert themselves to avoid such worst-case outcomes. Here and elsewhere, we draw morally indispensable distinctions between instrument and intent. Thus, when we analyze the privacy-eroding applications of computing and other information technologies, we hardly expect to abandon our critical judgment at the door.

This fantastical image of overbearing Technology—one can almost hear the capital T—has just the modicum of truth needed to slip under the radar of many people's critical faculties. Certainly, many outstandingly privacy-unfriendly practices would be impossible without recent strides in information technology. Think of the stunning sophistication of America's consumer credit reporting industry—tracking the lives and situations of hundreds of millions of consumers, attributing to each a three-digit score capturing his or her desirability as a credit risk, and adjusting these scores in real time, in light of the steady flow of personal information. Government tracking systems are no less astounding. Think of the sweeping (and widely successful) ambitions of the National Security Agency to monitor all Americans' telecommunications: all phone calls, emails, and web searches, all the time. Or the sophisticated abilities of the Department of Homeland Security to track the movements of both domestic and foreign air travelers—and to flag the movements of those deemed as warranting further attention to relevant agencies within the government. For the relentless growth in capabilities of spotting, tracking, and analyzing the lives of Americans, steady refinements in the technologies of surveillance are surely necessary conditions.

But they are not sufficient. In the social role of technology, sponsorship—what interests stand to flourish from its growth, and which ones stand to lose—matters decisively. To explain today's pervasive pressures on privacy, we need to focus on something more concrete than a disembodied mystique of Technology—and something much more familiar. That something is the role of well-organized, resourceful institutions. I mean here government and corporate organizations devoted to making and enforcing critical distinctions. Their work consists of discrimination in managing the affairs of large numbers of people.

What I mean by *discrimination* here is not necessarily *unfair* discrimination on the basis of race or religion, as the term is commonly used in American discourse. I mean all sorts of routine decision-making on how to treat people on the basis of their records—for example, discrimination in college admissions between applicants bracketed merely as "qualified" and those classified as "must admit," or discrimination between applicants judged as qualified or unqualified for financing to purchase a home. Any formal decision process may prove to be more or less just in the judgments it renders. But whatever the case with any particular such system, we live in a period when key steps in life are demarcated by some kind of formal discrimination at almost every turn.

In the world of business, the credit reporting industry shoulders much such work, tracking our consumption lives and ranking our desirability as credit customers. The same holds for marketing and advertising companies, devoted to determining which consumers are susceptible to which appeals for which sorts of products and services—and capable of paying for them. In government, such precise judgments of individual cases support assessment and collection of taxes, law enforcement, vehicle and driver licensing, controls over foreign and domestic travel, counterterrorism, and a host of other familiar activities. In all these connections, action requires recourse to people's *records*—systematically compiled results of tracking processes that follow us through our lives. This work of discrimination and enforcement thus generates both the ubiquitous *demands for personal information* in the twenty-first century and much of its *supply*. Let us call mechanisms for carrying out such discriminating enforcement *personal-decision systems*.

We do not necessarily experience the attentions of such systems as harsh or unfriendly. Think of drivers seeking to take to the road, travelers preparing to board domestic and international flights, consumers seeking credit cards or mortgages, even (in this author's case) a new city resident seeking municipal garbage-collection services. To institute such new activities or relationships, someone must connect our present situation with our "records." If frustrated in our wishes at such points, we ourselves grant a measure of legitimacy to these same processes by citing our own records as warranting better treatment. But, whether benign or coercive, personal-decision systems must necessarily apply some critical judgment process to the people they deal with—if only to determine that those seeking things like health care access, consumer credit, or tax refunds are indeed who they claim to be. These efforts to link each individual with the demonstrable facts of his or her case may seem routine and unavoidable, even for agencies committed to providing life-giving services. But no one could claim that they are conducive to privacy.

Sometimes the steps forging the crucial connections between a living person and the associated biography are strikingly unintuitive. Among these are systems based on technologies for computerized identification of faces, voices, vehicle license plate numbers, the locations of cell phones (in order to track the movements of their owners), or the DNA left behind on objects from cigarette butts to semen. Reliance on any of these and countless other forms of identification—eagerly developed and promoted by one private or corporate organization or another—changes the equation of who-can-know-what-about-whom. And such changes ultimately have their repercussions in new forms of authority and coordination. They also create new "maps" of the social world, as boundaries between domains of anonymity and those made up of actors whose histories, and hence whose future potentials, are known to those around them. They may create new opportunities for manipulation—for example, by identifying people as vulnerable to unjust disadvantage because of their identification as gay, as transgender, or as a former convict. The world is a more comfortable place, for many purposes, when people are taken at face value as they present themselves. Thus, we should think twice about tearing down the partial veils of ignorance that may sustain smooth social relations.

In a *New York Times* article, journalist Kashmir Hill told of an extraordinary encounter in an upscale New York Italian restaurant. Across a crowded dining room John Catsimatidis, the billionaire owner of Gristedes, a chain of high-end grocery stores, spotted his daughter being seated with a date unknown to him.[1] The magnate asked the waiter to snap a photo of the couple and then ran a quick search of the date's face on facial-recognition software that Catsimatidis himself had helped sponsor. The software, dubbed Clearview by its developers, immediately called up not only the man's name but many details of his biography.

Perhaps luckily for the young couple, Catsimatidis found nothing unacceptable in his daughter's choice of a dinner companion. But the same could not be said for some shoppers at one of his markets, where Clearview software had been field tested. According to the *Times* account, it identified "shoplifters or people who had held up other stores." Kashmir Hill goes on to report that the developers of the software had, by early 2020, sold their product to a labor union, a real estate firm, and "Best Buy, Macy's, the National Basketball Association and numerous other organizations."[2]

For public consumption, Clearview's creator, the Australian entrepreneur Hoan Ton-That, plays down the revolutionary implications of his invention. The services of his company, he contends, really don't do anything other than replicate the work of other search engines. Rather than making it possible to search for people's images by name, it reverses the process by linking their biographies to their images.

From this account, we do not know what interests the information produced by Clearview will serve—that is, what decision-making and enforcement activities will be based on the discrimination it affords. Probably there will be many would-be buyers. They could include political strategists (like the users of geofencing, described in the introduction) who want to attract select voters to their campaign rallies—or exclude those committed to the other side; law enforcement agents trying to detect terrorist sympathizers—or perhaps merely individuals deemed likely to develop such tendencies in the future. Other future users might be government agencies seeking to identify wanted persons in crowds at public events or in airports.

The possibilities appear endless—and we can be sure that the mere existence of this technology will inspire much original thinking on what new forms of discrimination it may support. After all, the data that make up the stock-in-trade of this system—people's facial features and the details of their biographies—are scarcely secret. Putting them together will amount to a far-reaching transformation in human relations. But do we really want to live in a world where the authorities—from the local police to the security specialists at American Express to the Department of Homeland Security—not only recognize us on sight, but can also instantly link such "hits" to the totality of our records?

True, the most important uses of these new powers cannot be predicted precisely. But no one who contends—as Ton-That does—that a revolutionary system like Clearview raises no new issues of morality or public philosophy deserves to be taken seriously. The initial intake that includes someone in the Clearview database may be painless and unobtrusive. But the long-term consequences of being included are apt to be profound.

A Social Invention

Institutions that accomplish such discrimination on a large scale represent a distinctive, if little-acknowledged, social invention of the mid-twentieth century. Considered as industries, they are producers of bases for "correct" decisions on literally millions of people every day—along with forceful action to implement such decisions. As Shoshana Zuboff has noted, the discriminations thus afforded offer organizations priceless insight into what people are likely to do, or want to do, next.[3] When organizations succeed in this quest, they create a basic requirement of social order—a world where the "right" people gain access to roles and privileges they are held to deserve, while the wrong ones are excluded or penalized. Without such discrimination, we would supposedly find ourselves living under regimes where criminals are inducted into positions of trust, profligate spenders are accorded sweeping credit privileges, revenues go uncollected as scofflaws thumb their noses at tax collectors, and on and on.

And if discriminating decision-making and forceful implementation are the essential outputs, then detailed, pertinent information on the people involved is the indispensable raw material.

Note a significant characteristic of personal-decision systems *in general*: their work is rarely complete; the discriminations they achieve by parsing people's records could almost always be finer. No credit reporting system perfectly predicts which consumers will pay and who will default; no terrorist-tracking system correctly identifies all those bent on violence; no tax compliance system perfectly targets those underreporting their incomes, and only them. Hence the constant quest for more, and more revealing, personal information. Like carnivores in the wild, most institutions of surveillance are, in some sense, always hungry for more personal data.

Of course, not all technologically supported pressures on privacy involve large institutions. Products like "big ear" listening devices that enable private parties to eavesdrop on distant, private conversations cause many one-on-one privacy breaches. So may DNA-based ancestry research or old-fashioned hacking of others' communications. But such practices are dwarfed in their net impact by pressures on privacy from government and private-sector institutions—both in their *demands* for personal data and in their *provision* of such data to satisfy demands from other parties.

The sheer impact of these systems on people's lives makes it obvious why they have triggered such widespread debate and, ultimately, regulation. Personal-decision systems seem to govern access to many of the good things in life—as well as guiding the application of many things almost everyone would prefer to avoid, like criminal identity or status as a poor credit risk. Accordingly, conflict between the institutions and the individuals involved is always at least latent. Keepers of personal data tend to demand more and more of it, to support more telling and precise decisions on individuals. The latter may push back at the data-keepers, insisting that certain information is too private to disclose, or simply that it is no one's business but their own.

These predictable tensions, in turn, fuel fierce policy debates. Should insurance companies be permitted to rate applicants for insurance coverage on the basis of their credit scores (a common American practice)? Should law enforcement agencies be authorized to retain DNA information they collect in investigations that result in no conviction? Should police have the option to track the movements of cell phone users via "stingray" technology without court orders (another widespread practice)? Should the IRS be permitted to draw credit reports to determine whether taxpayers' reported incomes correspond with their consumption habits? Should advertisers or government agencies be permitted to compile information on people's sexual orientations *at all?* Who is responsible when errors in people's files cause destructive mistakes in the ways they are treated? Questions like these, reflecting endemic conflicts of interest between the watchers and the watched, have surrounded personal-decision systems from their start.

Beginning in the 1960s, debate over such matters gradually led to redefinition of personal-decision systems as *public issues*—matters for legitimate concern and criticism, and ultimately for state intervention. The emerging consensus was that systems like these are too important to be left solely to the discretion of the organizations that create them. The result has been sustained, worldwide proliferation of national *privacy codes* and *personal data protection laws*—bodies of statutory law, court precedents, and policy directions intended to subject personal-decision systems to guidance in the public interest. The first of these date to the 1970s: in the United States, the Fair Credit Reporting Act (1970) and the Privacy Act (1974). In Europe, Sweden (1973), Germany (1977), France (1978), and Norway (1978) adopted broader privacy codes governing both private-sector and government systems. The spread of such adoptions has continued up to the present. At the time of this writing, more two hundred countries around the globe have created their own national codes devoted to protecting privacy—or *personal data protection,* to use the now-standard term. Among the more recent countries to join this global "privacy club" are Argentina (2001), Malaysia (2013), South Africa (2014), Mexico (2017), and Egypt (2020).

Privacy Codes

Despite many differences, these national personal data protection codes show marked "family resemblances." Some lay down criteria that must be met before any such system can legally operate. Most require that personal-data systems publicly identify themselves and their activities. Most afford the subjects of record systems access to at least some of their own records. Many make it possible to challenge specific entries in one's own files and specific decisions based on those files. Many restrict the sharing or release of filed data and specify uses that may be made of such data. And the majority of these codes—but not U.S. privacy law—establish a commissioner of personal data protection, backed by a (usually small) agency, endowed with varying degrees of independence. These offices are normally empowered to investigate privacy complaints and act as privacy advocates in the public forum. With some exceptions, these powers apply much less forcefully to law enforcement and national security agencies than to other personal-data-keeping parties.

Beyond this, national data protection codes vary enormously in strength and focus. Some create data protection commissioners with considerable independence from the executive branch. Elsewhere, the head of data protection is subordinate to the prime minister or another officeholder. Some codes endow the commissioner with considerable powers to investigate and prosecute offenders; others provide only for advisory powers. Some codes restrict the purposes for which personal-decision systems can be created in the first place; others do not.

The European Union has done better than most countries in this respect.[4] There, recent legislation has strengthened the European Community's original Privacy Directive of 1995 into today's General Data Protection Regulation, mercifully known as the GDPR. The new regulation effectively creates a single personal-data code for the entire European Union. It provides for data protection commissions and fully independent commissioners in each country, plus a centralized body for the entire EU. Unlike many of the world's privacy codes, that of the European Union applies only to private-sector organizations. It requires that every personal-data system have a "legal basis" to operate—so that the mere desire of any party to create such a system does not warrant permission to do so. It limits what can be done with personal data held in file, blocking most sharing of data, absent either a legal requirement to share or consent from the subject. And it grants to the data protection commissions that it establishes strong powers of investigation and enforcement over private-sector institutions, including the ability to impose fines on companies of up to 4 percent of their annual global turnover.

Most significantly, EU law defines a new right—the right to protection of one's personal data held in automated systems of any sort. This has far-reaching implications. It means, for example, that its guarantees extend even to forms of personal-data-keeping unknown at the time the laws were framed—newly invented systems are as much subject to the law as the familiar bank records, insurance systems, social security files, tax records, or any other long-established database. Its status as an individual right places personal data protection alongside freedom of expression or of association—not as an *absolute* good, certainly, but as a consideration worthy of weighing against such competing values as government efficiency and national security.

The language of the 2016 EU Regulation on the "processing of personal data and . . . the free movement of personal data," like the 1995 Privacy Directive that it replaced, emphasizes the intention to create a strong and uniform personal-data environment throughout Europe. The idea is to streamline economic processes within the European Union by affording the transfer of personal data among member countries without interference from inconsistent privacy rules. Transfers of personal data to third countries outside the European Union are permissible, provided only that the recipient country's data protection laws be judged "adequate" in relation to Europe's standards. To date, the European Union has recognized some twelve other countries, from Andorra to Uruguay, as providing adequate levels of protection to warrant such mutual sharing.

The United States has never been among those so designated. The options that American law offers for limiting dissemination of information on oneself are conspicuously weak, compared to EU protections. This failure to meet EU standards of adequacy has threatened to prove costly for many American corporations, who find their access to data on European employees and consumers blocked. The resulting pressures have twice moved officials on both sides of the Atlantic to paper over the conflict with arrangements to bypass the restrictions through special provisions aimed at protecting corporate transfers of personal data. The first of these, the "Safe Harbor" agreement, was in effect from 2000 to 2015, when it was struck down by the Court of Justice of the European Union, Europe's highest court. Edward Snowden, whose 2013 disclosures opened wide-ranging public debate over U.S. government surveillance of this country's "private" citizens, declared on Twitter that the action had "changed the world for the better." By contrast, the U.S. commerce secretary, Penny Pritzker, insisted that it "puts at risk the thriving trans-Atlantic digital economy."[5] Its replacement, the "Privacy Shield" agreement, was in effect from 2016 to 2020, when the same court struck it down as well.[6] Both decisions cited failure of the relevant agreement to establish in the United States key protections found in EU law.

A Privacy-Rights-Free Zone

The differences between European and U.S. law are dramatic—as much in their underlying logic as in their practical consequences for people's control over information about themselves. U.S. law, unlike its rights-based European counterpart, creates specific policy fixes as add-ons to existing data systems, setting down mechanisms for things like access to one's own files that vary from one form of recordkeeping to the next. Thus, in addition to the Fair Credit Reporting Act of 1970 and the Privacy Act of 1974, America has separate legislation governing financial transactions (1978), video rentals (1988), medical records (2000), and direct marketing to children (1999), among several other forms of recordkeeping. Personal-information systems *not* covered by such "sectoral" legislation operate largely free of privacy constraints. In America's private sector, at least, personal data often simply represent another business asset. Thus, client records of a clinical psychologist's practice, however confidentially held at the outset, can be subject to sale, along with furniture and other assets, upon the death or retirement of the psychologist.

Thanks to elastic interpretations of the Privacy Act of 1974, controls over sharing of data among U.S. government agencies are only slightly better. Yet European law regarding personal-data systems in the private sector forbids "secondary release" of such information for purposes other than those for which it was originally collected. In the United States, some sectoral legislation limits (though it hardly proscribes) such release in specific settings—consumers' choices of video rentals, for example. But most private-sector organizations, most of the time, have a free hand in selling, trading, or bartering data from personal-decision systems they maintain—whether those data have been provided by the person in question or "harvested" from the internet, public records, social media postings, or elsewhere. Recent changes in California law, discussed in chapter 7, mark a hopeful departure from this pattern.

This virtually free flow of personal data has spawned industries devoted to collecting such information by wholesale purchase and trade, and retailing it in the form of made-to-order reports. The purposes of these sales range from targeting consumers for unsolicited ads, to vetting prospective employees or business partners, to inquiries on potential dates, to selection of jury members by trial attorneys. Since the subjects of the reports are not parties to the transactions, most probably never know of the reporting. Consider the capabilities of Acxiom, one of America's most established data brokerages: "Today, Acxiom maintains its own database on about 190 million individuals and 126 million households in the United States. Separately, it manages customer databases for or works with 47 of the Fortune 100 companies. It also worked with the government after the September 2001 terrorist attacks, providing information about 11 of the 19 hijackers."[7]

Data brokerages maintain curious relationships with government data systems. Federal agencies are largely constrained by the Privacy Act of 1974 from releasing personal data held in their files to private-sector parties. But these agencies are also often reliable *consumers* of reports from private data brokerages. A study done in 2005 found that federal agencies spent approximately $30 million to purchase such reports, with law enforcement agencies accounting for 69 percent of the purchases and counterterrorism agencies generating 22 percent.[8] At the same time (as detailed in chapter 2), other agencies of state, county, and local government provide data on such revealing matters as tax payments and liens, vehicle registration data and driving records, court proceedings, residential history, and much else. All such data are apt to be incorporated in the files of data brokers.

In Europe, companies compiling and selling such information would have to acknowledge these activities publicly—and grant those listed the option of removing themselves from the database. The United States has no such categorical requirement. As every American knows, junk mail, spam, and other appeals originating from uncontrolled exchanges of mailing lists, email addresses, and phone numbers represent the informational equivalent of crabgrass in everyday life. Efforts to determine the origins of such intrusions are mostly doomed to failure. In the United States, under federal law, doing business as a seller of personal information carries no obligation to acknowledge one's activities to the public or to reveal to anyone whether their data have been disclosed. The "finders-keepers" principle continues to dominate; if personal data are perceived as "public" at any point, how can it be wrong to repackage and retail such data?

But ultimately, a public philosophy like this devolves into cynical practices. Consider the story of Richard Guthrie, a ninety-two-year-old U.S. Army veteran and a widower, targeted by professional data merchants who bought his information in a market whose very existence he was probably unaware of. "Guthrie, who lives in Iowa," reported Charles Duhigg in the *New York Times*, "had entered a sweepstakes that caused his name to appear in a database" that was then advertised by InfoUSA, "one of the largest compilers of consumer information. InfoUSA sold his name, and data on other elderly Americans, to known lawbreakers, regulators say." InfoUSA advertised lists of "Elderly Opportunity Seekers"—3.3 million people "looking for ways to make money"; "Suffering Seniors"—4.7 million who had cancer or Alzheimer's disease; and "Oldies but Goodies"—five hundred thousand gamblers older than fifty-five. Offering the individuals' information for 8.5 cents apiece, the company said about one of these lists, "These people are gullible. They want to believe that their luck can change." Guthrie's "telephone rang day and night. After criminals tricked him into disclosing his banking information, they . . . raided his account." The result was heartbreaking. "I loved getting those calls," he told an interviewer. "Since my wife passed away, I don't have many people to talk with. I didn't even know they were stealing from me, until everything was gone."[9]

What happened to Richard Guthrie is intolerable. Surely even those who see no *necessary* harm in the sheer compilation of personal information should recognize the risks of operations that make such targeting and exploitation possible. What InfoUSA sold to the parties who swindled Guthrie was his data—but not *just* his data. It was a road map of his life situation, including his vulnerabilities. He was no match for the forces arrayed against him. It is impossible to see what higher value justifies collection and retailing of data for uses with such devastating effects.

American Privacy Grumblings

For many, the privacy-rights-free zone in American law actually represents liberated territory. The internet age, these observers hold, has inspired Americans to abandon anachronistic privacy expectations. The evidence shows, they say, that Americans readily jettison their control over information about themselves in exchange for conveniences and creature comforts. "Just note how blithely Americans 'click through' the terms of service posted by website and app sponsors," these observers are apt to remind us, "or how little attention they pay to privacy notices on websites or software packages." Such actions are nothing less than rational, in this view, considering the obvious benefits of such things as the "free" products and services obtained in exchange for our data.

That many of us bow to these offers-we-can't-seem-to-refuse is beyond doubt. The question is, should *acquiescence* stand as evidence of support for wholesale loss of control over our data? In fact, abundant evidence attests to Americans' dissatisfactions with the free flow of data about themselves. In 2014, when the respected Pew Research Center asked a representative national sample of respondents about the level of importance they ascribed to having control over what information is collected about them, 90 percent responded that it is either "very important" (65%) or "somewhat important" (25%). Approaching the issue in slightly different terms, the Pew researchers also asked respondents to rate the importance of "being in control of *who* can get information about them," to which 93 percent replied either "very important" (74%) or "somewhat important" (19%). Another Pew survey conducted in 2014 sought respondents' views on the statement that "consumers have lost control over how personal information is collected and used by companies"; 91 percent expressed either strong agreement (45%) or agreement (46%).[10]

Despite such findings, many commentators have ascribed to American consumers what they term "pragmatic" attitudes toward appropriation of their personal information—that is, a willingness to consider significant loss of privacy as little more than fair exchange for the advantages and conveniences that supposedly result.[11] Such interpretations strike me more as wishful thinking on behalf of America's personal data-using industries. There is a sharp difference, after all, between informed consent and uninformed acquiescence. Some ingenious research by Chris Jay Hoofnagle and Jennifer Urban, conducted at about the same time as the Pew studies cited above, reveals confusion and ignorance among consumers as to what exactly they are giving up in their online transactions. Hoofnagle and Urban agree that some sentiments of consumers could be characterized as "pragmatic." Yet when they dug more deeply into their respondents' verifiable knowledge of the movements of and constraints on data about themselves, they found limited sophistication. Even the so-called pragmatists, confronted with examples drawn from real-world experience, rejected the exchanges and disclosures that pragmatists were supposed to accept.[12]

In a careful telephone study of some 1,506 Americans, Joseph Turow and collaborators found little evidence of any "social contract" between consumers and those who appropriated their personal data.[13] "Rather than feeling able to make choices," the authors write, "Americans believe it is futile to manage what companies can learn about them."

. . .

Skeptics often dismiss such findings as mere talk. After all, they hold, what significance can one ascribe to statements that consumers assented to in surveys that appear to be blatantly inconsistent with people's actions? This complaint has been widespread enough to receive its own name, "the privacy paradox"—a sweeping reference to the alleged discrepancies between open expression of privacy concerns and public actions to address such concerns. Long-running debate over this so-called paradox moved my student collaborators and me to create the chart in appendix 2, a visual representation of "privacy setbacks" and "privacy affirmations" during the past four decades. We found that both privacy setbacks and privacy affirmations are authentic and varied. The affirmations take very different forms, some involving hundreds of thousands of people, such as demonstrations in Australia in the 1980s, against institution of a mandatory national ID card. But the chart also includes a mobilization of parents of schoolchildren in the small country town of Sutter, California, in 2005 that struck us as particularly dramatic in context. There, aggrieved parents rose up—successfully—against a proposed requirement for a computerized tracking device to be worn by all pupils in the local elementary school while on campus. "Our children are not inventory" was the rallying cry of the indignant parents.

Perhaps the privacy affirmation with the greatest potential for long-term consequence among the cases gathered in appendix 2 is the mobilization leading to passage of the California Consumer Privacy Act of 2018. It grants California consumers the following rights when dealing with large personal-data-consuming organizations: to know what data about them are being collected; to know whether such data are sold or disclosed, and to whom; to decline the sale of such data; to access such data; to request deletion of such data; and not to be discriminated against for exercising these rights. Many of these rights closely parallel those sought in the reform proposals put forward in this book. The law is just going into effect at the time of this writing, but if these new rights withstand challenge, the implications for privacy throughout the country could be sweeping.

Given the chance, then, publics around the world have acted to express their objections to what they regarded as high-handed uses of their data. The problem is that activists have managed to politicize so few of these "teachable moments" where loss of privacy is all but dramatized before our eyes—for example, the one-sided agreements one is expected to sign in order to access even minimal services over the internet. Virtually every system for appropriating and taking advantage of personal data in this country bears the "fingerprints" of one or another resourceful organization, public or private. Few if any of these organizational sponsors want to see those fingerprints erased. Thus, legislation that established the federal Do Not Call List explicitly exempted members of Congress from its provisions, leaving them free to bombard prospective voters with robocalls that would be illegal if other parties commissioned them. Our elected representatives had the political sense not to loosen their own grip on our eardrums.

Innovation Politics

The forces arrayed against strong privacy are never so mysterious as the vague "imperatives" often ascribed to Technology. What both the privacy setbacks and affirmations depicted in appendix 2 reflect is simply the oldest of stories—political-tug-of-war between partisans of opposing interests. On one side, we have representatives of resourceful and well-organized institutions, bent on enlarging their place in the sun by doing more of what they have already learned to do very well. On the other, we have embattled grassroots groups campaigning for broader individual choice over the fate of "their" data—sometimes successfully, sometimes not. The outcomes of these struggles depend on such earthy and familiar dynamics as the state of public opinion, the role of professional lobbyists, the organization of grassroots efforts, and the dramaturgy of public events—personal-data scandals, shocking privacy horror stories, and the like. Once such battles are won or lost, we can confidently expect purveyors of technology to step forward, ready to implement prevailing directions—whether these involve tracking people more closely or providing the latter with better tools to control use of their data.

But let us be frank. American privacy interests have not seen many signal victories in these battles. Appendix 2 notes a number of privacy affirmations emanating from the United States. But what it counts as such are episodes of significant *mobilization* of grassroots involvement, not necessarily success in promoting or defeating policies. This country has inspired some of the most penetrating examples of research and writing that warn of the dangers of overweening surveillance. Yet the evolution of personal-decision systems in the ensuing years has driven practice steadily in the opposite direction. The strategic advantages to data-keeping organizations of exploiting personal information gathered in one context to buttress decision-making elsewhere have simply proven irresistible—as much in government agencies as in corporate recordkeeping. Symbiosis and exchange of personal data across systems have grown to be more the rule than the exception—as in the wholesale merchandising of credit scores for use in screening applications for insurance coverage. With a keen eye to new commercial opportunities, the consumer credit reporting industry has become a major source of personal data for retail marketing, despite regulatory efforts to limit its activities to producing data for evaluation of credit applications.

Thus, businesses that acquire vast amounts of personal data when screening consumers' credit applications have learned to resell or reuse such data to support campaigns targeting the same customers for advertising appeals.[14] Elsewhere in the private sector, entire industries have grown up that are devoted to treating personal information as a commodity, for sale or trade without the consent or even the knowledge of the subjects (chapter 3 examines this trend at greater length).

In government recordkeeping, elastic interpretations of the Privacy Act of 1974 and other legal victories for pro-surveillance forces have encouraged wide sharing of such information among agencies—to say nothing of the unseen and unfettered exchange among national security and law enforcement agencies that is largely beyond the reach of privacy legislation. As a result, Americans seem to have embraced the assumption that, once their information is disclosed within one setting, it is apt to surface virtually anywhere else—often serving quite different purposes and interests. This free flow of personal data is widely touted as an immutable requirement of Technology, or an expression of the ethos of the "Information Society." But in the European Union, the strict legal ban on "secondary release" of filed data—that is, its disclosure for new purposes—creates a far more private world. The propensity to exploit available data for new purposes is hardly a direct effect of Technology; rather, it is a reflection of the balance of power among contending interests. Not to be too subtle, there is no reason why Americans concerned about their privacy shouldn't see this desire satisfied—except for the influence of major institutions that have grown up to profit from its capture and sale.

But any idea of curtailing the "free flow" of personal information, the anti-privacy voices will warn, could threaten Innovation—the first cousin, apparently, of Technology. Just think, they will insist, of the cornucopia of services and conveniences that we owe to the freedom of (mostly American) companies to experiment with personal information. Think of the new ideas, the great industries, and the monumental new American fortunes built in a world where personal information is available without legal obstacles or great expense. Like fresh air—back when there was plenty of it! The systems that create these "free" products purchased unobtrusively through the harvest of one's data, enthusiasts hold, have produced the golden eggs that have made this country the world leader in Innovation. Without the prospect of *monetizing* these new accumulations of personal information to encourage further experimentation and risk taking, all Americans would be poorer. From this point of view, proposals for a right to "just say no" to the appropriation of one's own data make about as much sense as suggestions for steak tartare as the main course at a vegan banquet.

American Privacy Activism

Against these forces, some impressive activist groups have risen to defend Americans' privacy interests. In Washington, EPIC (the Electronic Privacy Information Center), the Center for Democracy and Technology, the American Civil Liberties Union, and other groups raise their voices, and lobbying skills, on behalf of privacy values. Their representatives testify in Congress; hold news conferences to decry privacy-invading innovations; or litigate privacy issues in court. Outside the capital, groups like Privacy Rights Clearinghouse in San Diego and the Electronic Frontier Foundation in San Francisco pursue a similar agenda—counseling consumers on privacy troubles in the first case, battling in court against privacy-eroding practices in the second. Like other public-interest lobbying groups, they have won a reputation for "punching above their weight"—obtaining results, like legal requirements for disclosure of personal data breaches, comparable to those achieved by lobbies for far more established, better-resourced interests.

The activities of these high-energy privacy activists evoke the work of dedicated but hard-pressed volunteer firefighters. They race from one privacy conflagration to the next, with scarcely a chance to draw breath before the next emergency. Yet these and other inspiring organizations have not yet succeeded in developing a major constituency for a grassroots privacy movement. One crucial missing ingredient, I hold, is a persuasive, comprehensive *vision* of a more private world—a vision strong enough to attract prospective supporters. Such a vision would convey an idea of how American life could work with far fewer demands on personal information and far more options as to when to share one's information, and when and how to resist such sharing. It would convincingly show how to protect citizens and consumers from surveillance and forced disclosures that are both hurtful to individuals and subversive of shared values of openness and pluralism in thought and action. It would mean putting forward new legal principles offering clear improvements over existing American privacy codes in the ongoing struggles between ordinary citizens and data-hungry organizations.

• • •

Unfortunately, when Americans do encounter privacy protection law, it is often in forms calculated to make privacy synonymous with meddling and frustration. We all run up against these deadening reminders of our privacy rights in the form of irksome advisories and tedious texts interposed between us and whatever we need to do next. I am thinking of website privacy notices, for example, or notices of terms of use like those on computerized services. It's not just that the language of these communications is tedious and anesthetic to critical thought. It's that the interruption comes just at the crucial moment when the user is focused on accomplishing something else. To the delight of the service providers, we impatiently click through these often-impenetrable gatekeeping disclaimers, resolving to worry about privacy another day, if ever. The net result is to convey the message that privacy protection is something designed to stop you from getting on with your life. This is not the association that privacy activists ought to be seeking.

In similar ways, privacy advocates have often allowed their concerns to be portrayed as highly technical specialisms that ordinary mortals cannot expect to penetrate. To a degree, this is unavoidable; upholding privacy often requires special expertise. But when building a movement, charting a route that leads through conceptual quagmire is a prescription for defeat. Even the most privacy-friendly members of the public may find it hard to stay interested in ideas like "privacy dashboards" or the details of cross-border data flows. These are indispensable concerns for professional privacy enforcers, to be sure. But we cannot afford to allow them to remain the main public face of privacy. Instead, we need a hard-hitting, simple program for sweeping improvements in Americans' relations with personal-decision systems and the organizations that deploy them. In other words, those determined to take privacy seriously need to approach the public with their own programmatic manifesto.

No social movement can succeed without projecting such a simple and compelling vision of a better world—a vision where disturbing realities of life-as-it-is yield to a believable new order that is clearly preferable. Builders of movements accordingly go to great lengths to create such visions. Some even manage to incorporate their alternative worlds in their names—Zero Population Growth; Right to Life; Marriage Equality; the Trust for Public Land; or even (in the United Kingdom) Frack Off.

Asked in 1915 what labor wanted, pioneering union leader Samuel Gompers is remembered for replying, simply, "More." In fact, though, he went on eloquently and at length: "more schools and fewer jails, more books and less arsenals, more justice and less revenge." Privacy activists need to project their own list of specific changes (as Gompers did), measures with transparent value to a broad public. The sought-after changes must impress would-be supporters at once as credible, workable, and worth striving for. And, most privacy advocates would insist, they must address the perceived privacy needs of the broad middle ranges of public opinion, while also coming to the defense of those most grievously hurt by the system's inequities.

Toward Strong Privacy

American privacy advocacy has produced few if any such programs. Privacy activists apparently find it easier to decry what they are *against* than to uphold a comprehensive alternative vision. Who wouldn't, given the challenges involved? Any attempt to sketch the outlines of a world where treatment of personal information actually fulfills its highest potential confronts us with the hardest questions—those about what we ultimately want.

Privacy advocates readily identify themselves as favoring more individual control over one's own information, for example. Yet no privacy proponent known to me would propose *unlimited* control over dissemination of data on oneself. In our personal lives, any such principle would strip us of the elementary ability to form critical judgments about those around us—on whether to extend our trust to specific people, for example, or withhold it. In public life, the ability to censor one's own past would enable candidates and officeholders to suppress unseemly incidents or earlier positions from the public record—hardly a good thing for anyone's concept of democracy. Thus, those who speak for privacy have no choice but to weigh *how much* and *what kind* of control over personal information they are prepared to advocate.

That means grappling with questions like these: What constitutes adequate grounds for creating and maintaining any system of personal information in the first place? Should First Amendment rights guarantee anyone the option of creating a personal-decision system for any purpose of his or her choice—as is substantially the case in this country today? Or should such systems only be permitted for socially sanctioned purposes—as in the requirement for "legal bases" for personal-data systems that is central to European law? Should the proclaimed "needs" of government and corporate organizations for more and more personal information be accepted at face value? Or should we expect law and policy to distinguish which "needs" deserve protection in this connection? Who "owns" the right to buy and sell personal data for profit-seeking purposes? Are there any forms of personal information that should simply never be marketed?

As we confront such questions, let us never imagine that we need only follow our gut instincts to draw a bright line between acceptable and unacceptable uses of personal information. No doubt we all have such a sense—bracketing what we consider unwarranted demands for personal data as many do pornography: "I know it when I see it." But as with pornography, gut reactions on privacy differ from person to person, and within each of us over time. Indeed, few sensibilities have proven themselves more malleable than notions of what one ought or ought not expect to disclose in the face of corporate or government demands for our personal information.

Along with differences over such questions of existential taste, another source of disagreement arises from questions of *feasibility*—over how measures and mechanisms proposed to make personal-decision systems as privacy-friendly as possible would work in practice. Debate on these matters is more likely to be informed by empirical evidence—and hence more likely to be settled. A proposal to protect privacy by requiring every citizen and consumer to maintain a listing of all files on himself or herself, for example, would surely demand more of the public than anyone could reasonably expect. Similarly, some early privacy advocates proposed that every personal-decision system be required to dispatch a copy of everyone's record to the subject of that record every year. But informed observers concluded that such a measure would be expensive and onerous to carry out—as well as being self-defeating, by disseminating so many copies of personal files without a certain destination. The keepers of many personal-data systems would often have to guess where to send the data, given the difficulty of keeping mailing and email addresses current. In cases like these, empirically based debate over what might or might not work in practice has at least a chance of narrowing differences of opinion.

In short, proposing any positive vision for a better regime of personal information involves a complex marriage of deeply held moral "tastes" and assessments of what practices are most likely to work under specified conditions to serve such tastes. The program for reform that I present here is no exception. The best that I—or any advocate—can do is to avow frankly, in advance, the bases of his or her positions. Thus, the proposals put forward here derive in large measure from a "taste" for stronger privacy, along with the following convictions:

1. Privacy—however difficult to define precisely—deserves the status of a *right*, on a par with other public goods deemed worthy of

legal defense. Comparable values might include freedom of expression, habeas corpus, or the right to own private property. Like other basic rights, privacy cannot be absolute, but must always remain in useful tension with other rights and claims. Taking privacy seriously means, by my lights, assuming that control over data on oneself must remain in one's own hands as the *default condition*. Thus, statutory and other requirements for disclosure require persuasive arguments to override the default. Privacy isn't everything, and many junctures demand that we subordinate privacy considerations to concerns for safety, freedom of expression, or needs to cope with emergencies like pandemics and other natural disasters. But privacy advocates expect the case to be made anew each time some sort of override is necessary.

2. Personal-decision systems—that is, routinized institutional uses of personal data to support decision-making on the people depicted in the data—must be accorded a special status in law and policy. Like the use of *controlled substances,* they must be recognized as often necessary, but also sufficiently destructive if abused, to warrant careful guidance and to require reasoned *legal bases* for their existence.

3. The fact that any organization has the resources to create or maintain a personal-decision system should not establish its right to operate such a system. Many other interests have to be taken into consideration, as well, including the privacy of individuals and the dangers to holistic values like freedom of expression, should the defense of privacy grow weak.

4. Efforts to protect information privacy have often failed, thus far, because of overreliance on institutional action to protect people's interests in their data. Whether the institutions are European-style personal data protection commissions or other bodies, the powers of the responsible authorities to monitor what happens to people's information have often been unequal to the task. With the relentless

increase in reliance on personal-decision systems, the expansion of personal-data-keeping has often outstripped the capacities of the appointed institutions. Accordingly, the reforms proposed here aim at encouraging and facilitating individuals' own efforts to inform themselves on the fate of their own data—and to take at least initial steps for enforcement on their own behalf. Many of the rights proposed here envisage direct appeals by individuals to data-keeping institutions—to invoke the right to resign from personal-decision systems not required by law, for example; or to require explanation of why personal information needs to be collected in the first place. The premise for these measures is that the institutional surveillance affecting ordinary "private" citizens today has become so pervasive that action from the principals has become the surest way of identifying failures to comply with basic privacy rights. This is why chapter 4 proposes the creation of two vast national portals, open to any American, for wide disclosure on all personal-decision systems.

5. Americans have become inured to signing away privacy options, implicitly or explicitly, in order to get on with their business on websites, internet transactions, software contracts, and the like. Taking privacy seriously requires establishing meaningful and durable protections that cannot be signed away so readily. This is best accomplished by establishing broad rights that apply to entire categories of situations and transactions, rather than options that must be renegotiated for each new relationship.

6. Decision-making that is *expedient* or *efficient* for organizations creating personal-decision systems is not necessarily *just* or *desirable* from the standpoint of the individuals concerned, or for the larger community. Justice and compassion for individuals, and prudent care to avoid creation of overweening central power, must often override efficiency concerns.

7. Law and policy on personal-decision systems must aim both at protecting individuals from personal harm *and* at preventing degradation of *public life as a whole*—for example, by creating chilling effects that may inhibit or degrade the quality of public life, even when individuals feel secure in their own privacy.

8. One ultimate aim of these reforms must always be to broaden the choices available to individuals regarding records on themselves—preferably by establishing simple steps that people can take on their own to enhance their privacy. In extremis, and with carefully delineated exceptions, people must retain the right to withdraw from most personal-data systems that are not legally required.

This list is by no means complete. But it does provide a basis for the program of reforms that structures the rest of this book:

1. Require every personal-decision system to have a legal basis—that is, a justification for its existence. To claim such a basis, the sponsors of every system must establish that its purposes support social values broader than just the interests of the sponsors.

2. Require consent from subjects of filed data for disclosure or sharing of their data, except as explicitly required by law. Forbid third-party release of such data, except as required by law.

3. Create a pair of national internet portals to afford Americans easy access to (1) complete information on the workings of personal-decision systems in this country in general; and (2) upon proper self-identification, the detail of one's own files, wherever they are located. Use information brought to light in these two sources to require deletion of outdated, inaccurate, or inappropriate personal data.

4. Create a "right to resign" from most personal-decision systems where participation is not legally required.

5. Create property rights over commercial exploitation of data on oneself, so that no buying, selling, or trade of such information is possible without consent of the subject.

The following five chapters develop a case for each of these objectives.

Of course, there will be many objections to these proposals, both opportunistic and sincere. Among the latter will be worries about dangers to First Amendment rights, including such basic democratic prerogatives as decrying the actions and character of one's political opponents. This and a host of other standard democratic practices, after all, depend on the ability to record information about people whose actions or characteristics quite properly draw critical attention from fellow citizens.

But note that these reforms aim only to constrain the use of personal data held in personal-decision systems, as defined in this book. They have no bearing on data compiled by journalists or writers in preparing news articles or opinion statements; on research done by political activists planning campaign appeals; or even on the activities of obsessive gossips bent on amassing accounts of scandalous actions of their neighbors. Only when sources of personal information are compiled in forms lending themselves to easy retrieval and systematic decision-making by organizations do they trigger the rights and restrictions proposed here.

All of this leaves much more to be said about implementation. How do we know what should count as a personal-decision system, for example? A few names and pertinent facts, jotted on the back of a bureaucrat's envelope? A secret White House memo listing Richard Nixon's political enemies? A list of bad debts from a decade ago?

None of the above, I would propose—if only because these particular examples do not involve a recurring system of decision-making and enforcement. But that leaves plenty of salient questions, with immediate implications for practice. Much key American privacy legislation, for example, relies on language like that used in the Privacy Act of 1974, which deals with "systems of records," defined as "a group of any records under the control of any agency from which information is retrieved by the name of the individual or by some identifying number, symbol, or other identifying particular assigned to that individual."[15]

One trouble here is that computing and data management have changed profoundly since 1974. Key decisions on individuals may now be made by drawing from bodies of data that are not part of any one "group" of records. Thus, officials with access to a variety of databases can readily search across many of them, possibly applying different identifying principles to draw together what could amount to a custom-made dossier on someone.[16] By the time these words reach print, one or more reformulations of what constitutes a "system of records" may be coming into use. Here as elsewhere in this book, I do not intend to tie the principles underlying these reforms too closely to specific technologies or techniques of data management. I expect that reforms proposed here will continue to represent alternatives to current practices, even as personal-data management practices evolve.

"Christianity has not been tried and found wanting," wrote G. K. Chesterton. "It has been found difficult and not tried." This is much my feeling about the privacy reforms proposed here. These ideas have long been broached, but only very reluctantly and unevenly tested in practice. The reason for this mixed success is not the imperatives of Technology or the inherent complexity of personal information processes, but old-fashioned political resistance. Privacy protection measures put forward thus far have simply not matched the magnitude of the forces arrayed against them. Any effort to loosen the grip of the opposing corporate and government actors threatens political strife of vast proportions. The reforms put forward here, if successful, would challenge the hold of the institutions involved on the people they deal with—inevitably eroding their dominant roles in our political and economic lives.

Thus, none of these ideas is altogether new. Some are already instated, in stronger or weaker form, in the law of the European Union or of other jurisdictions. Others have long been proposed and debated by privacy advocates. Yet their effect, if enacted together, would be synergistic: the various provisions would actually support each other, so that the whole would become more powerful than the sum of its parts. The *ensemble* would point the way to a markedly more privacy-friendly world—one where ordinary citizens and consumers have vastly more options over "their" data than is the case today. Moreover, these proposals aim at fundamentally reshaping the balance of power and initiative between resourceful institutions and the gravely underrepresented citizens confronting them. At several key points, they create for the latter the option to "just say no" to tracking by personal-decision systems. That is why I describe these proposals not so much as complicated to grasp, but as *excruciating* in the conflicts they portend.

Notes

1. Kashmir Hill, "The Secretive Company That Might End Privacy as We Know It," *New York Times,* Jan. 18, 2020.

2. Ibid.

3. Shoshana Zuboff, *The Age of Surveillance Capitalism* (New York: Public Affairs, 2019), 200–01.

4. Francesca Bignami, "Privacy and Law Enforcement in the European Union: The Data Retention Directive," *Chicago Journal of International Law* 8, no. 1 (2007): article 13.

5. Robert Levine, "Behind the European Privacy Ruling That's Confounding Silicon Valley," *New York Times,* Oct. 9, 2015.

6. Court of Justice of the European Union, "The Court of Justice invalidates Decision 2016/1250 on the adequacy of the protection provided by the EU-US

Data Protection Shield," July 16, 2020, https://curia.europa.eu/jcms/upload/docs/application/pdf/2020-07/cp200091en.pdf.

7. Natasha Singer, "Mapping, and Sharing, the Consumer Genome," *New York Times*, June 16, 2012.

8. Evan Hendricks, "Privacy Times," *Privacy Times* 26, no. 8 (April 2006), 1.

9. Charles Duhigg, "Data Miners Swindle the Lonely and Elderly," *New York Times*, May 21, 2007.

10. Mary Madden and Lee Rainie, "Americans' Attitudes about Privacy, Security and Surveillance," Pew Research Center, May 20, 2015, https://www.pewresearch.org/internet/2015/05/20/americans-attitudes-about-privacy-security-and-surveillance/; Mary Madden, "Public Perceptions of Privacy and Security in the Post-Snowden Era," Pew Research Center, November 12, 2014, https://www.pewresearch.org/internet/2014/11/12/public-privacy-perceptions/; see also Lee Rainie et al., "Anonymity, Privacy, and Security Online," Pew Research Center, September 5, 2013, https://www.pewresearch.org/internet/2013/09/05/anonymity-privacy-and-security-online.

11. Alan F. Westin, "Social and Political Dimensions of Privacy," *Journal of Social Issues* 59, no. 2 (2003): 431–53; see esp. 445–48.

12. Chris Jay Hoofnagle and Jennifer Urban, "Alan Westin's Privacy Homo economicus," *Wake Forest Law Review* 49, no. 2 (June 2014): 261–318, 288–89.

13. Joseph Turow, Michael Hennessy, and Nora A. Draper, "The Tradeoff Fallacy: How Marketers Are Misrepresenting American Consumers and Opening Them Up to Exploitation," Annenberg School for Communication, University of Pennsylvania, 2015.

14. Josh Lauer, "From Debts to Data," in *Creditworthy: A History of Consumer Surveillance and Financial Identity in America* (New York: Columbia University Press, 2017).

15. Privacy Act of 1974, Public Law 93-579, 5 U.S.C. 552A, *U.S. Statutes at Large* 88 (1974).

16. See Robert Gellman, "Report: From the Filing Cabinet to the Cloud: Updating the Privacy Act of 1974" (May 2021), https://www.worldprivacyforum.org/2021/05/from-the-filing-cabinet-to-the-cloud-updating-the-privacy-act-of-1974/.

2 Ban Personal-Decision Systems That Violate Core Values

"One more thing," the sixty-four-year-old patient told Dr. Janelle Duah, as his examination was coming to a close. "I've got blisters." As recounted in a 2016 *New York Times* article by Lisa Sanders, the tiny blisters on the man's hands did not correspond to any pattern familiar to the physician, who was working at a Veterans Administration hospital in West Haven, Connecticut. The blisters appeared irregularly, three or four times a month, and sometimes embarrassed the patient when they burst and ran. They had been occurring for twenty years, without a conclusive diagnosis.

Dr. Duah found these reports strange enough to warrant further investigation. The patient was generally healthy, active, and not greatly inconvenienced by the blisters. But there was something uncanny about this symptom. She turned to the VA's electronic medical record system, containing data from as far back as twenty years. The patient's record showed that in an earlier examination, the level of iron in his blood had been inordinately high—high enough that the examining physician had flagged the result, without following up. Further tests showed that the excess iron put the patient in serious danger of fatal repercussions, including cirrhosis of the liver and heart failure. He is now under the care of a hematologist for hemochromatosis, a rare but treatable disease caused by a genetic abnormality.[1]

Dr. Duah's patient has reason for gratitude. Unknowingly afflicted by a rare disease that could have killed him, he was lucky to be treated by a doctor willing to focus on symptoms that at first seemed merely curious. He was no less lucky to be seen in a hospital with computerized access to his detailed medical history, and vital intelligence on possible connections between his symptoms and other rare occurrences. By drawing together disparate strands of information from these sources, those treating him could pinpoint the intervention he needed before he grew seriously ill.

But not all encounters with personal-decision systems end on this upbeat note. Consider the following account of one Harlem resident's search for housing, from a 2016 *New York Times* article by Kim Barker and Jessica Silver-Greenberg:

> After two years of being homeless, napping in stores open all night and more recently staying in a convent in Harlem, Margot Miller found out in March that her luck was about to change: She had qualified for an apartment for low-income older adults.
>
> "This is to inform you that a rental unit has become available," the letter from the building's owner, Prince Hall Plaza, began.
>
> Elated, Ms. Miller, 68, said she immediately went to the building's office to claim the apartment. But after a background check, she said, the building reversed course.
>
> "I go there, I'm all excited," Ms. Miller said. "The woman there then does something on the computer. Then she said, 'You can't have this.'"
>
> She was disqualified, the woman told her. Not because of her credit score. (At 760, hers was stellar.) And not because of a criminal record. (She had none.)

The cause of Miller's dashed hopes was yet another data system—a privately owned and operated organization that tracks cases in New

York's housing court and is known as the "tenants blacklist." Miller had been sued by an earlier landlord, and a record of that suit had been retained in the blacklist. "'There are a lot of tenants who are terrified of complaining or of withholding rent because they are afraid of getting on these blacklists,' said James B. Fishman, a lawyer who has settled two class-action lawsuits against tenant-screening companies."[2]

Many people in Miller's position apparently have no understanding of the forces working against them. The *New York Times* writers note that

> in 2014, the [New York City] Council passed legislation requiring landlords, property managers and brokers to disclose which screening company, if any, they used for background checks. Tenants can order their files from the companies and correct inaccuracies.
>
> But for people who live on the margins, without a permanent mailing address, fighting the list can be difficult. Until they are denied housing, some tenants have no idea that they are on the list at all.[3]

Ironically, one could say that Dr. Duah's patient and Margot Miller had the same experience—an experience repeated countless times every day. They found themselves unexpectedly dealing with organizations bent on associating them with their pasts. Duah's patient was fortunate to find himself in a resourceful facility, dedicated to quick, informed intervention in defense of his life and health. Miller was unlucky in that the detail of her tug-of-war with a former landlord ended up in another system—this one created and administered on behalf of New York City's rental housing industry. The New York Housing Court had sold details of her case to companies specializing in reporting tenants' histories to prospective landlords. The latter recoil at dealing with tenants who have been in court over housing issues. The two personal-decision systems, very similar in structure, were clearly serving different parties and different interests.

Note that there is no reason to believe that Miller's debacle stemmed from a "mistake"—inclusion of inaccurate or biased information in her files, for example. In a tight housing market, property owners find it in their interest to reject any and all tenants with histories of legal tussles with landlords. Given intense demand for rental housing, any owner would be well advised to choose new tenants who showed no such rough edges.

This highlights a basic and sobering truth—not just about housing and health, but about personal-decision systems *in general*. Uses of personal information that are *efficient* or *expedient* for sponsors of such systems are by no means necessarily *just*. Justice in the treatment of individuals, including consideration of conflicting points of view, extenuating circumstances, and the like, is an expensive, labor-intensive, often inconclusive business. For managers of personal-decision systems, simply ignoring such niceties may offer much the most efficient route to success.

The same logic dominates workings of the most varied systems. Insurance companies, for example, exert themselves to exploit every scrap of available personal data to write policies on the most profitable possible terms. Not all of this crucial information has to be sought from outside sources. Acting strictly within their own narrow interests, insurance companies would be shrewd to raise their rates for customers who make frequent *inquiries* about possible claims—on the assumption that they are more likely than the average customer to put thoughts into action at some point in the future.

Such practices should hardly shock anyone. The companies are within their legal rights to take the fullest advantage of all personal information at their disposal. But if our aim is to use information technologies to support better, more privacy-friendly relations between organizations and the public, we should expect something better.

Fingerprints

An acute observer once characterized technologies as *prostheses* for human faculties—extensions of our innate physical and mental powers, like stilts or telescopes. Personal-decision systems fit this characterization particularly well. They enhance human abilities to gather, sift, and analyze information on human lives and make forceful decisions in response. Doing for resourceful institutions what handwritten notes and personal observation do for individuals, such systems confer sweeping new powers to control and direct. They accomplish this by enabling the organizations that create them to *reach down* into otherwise anonymous or uncooperative populations to *single out* persons of interest and to enforce *crucial decisions* about them, allocating to each *just* the "right" treatment, based on his or her record.

Some would consider this expanding role of authoritative information to be strictly good news. In an "information society," they profess, every decision becomes better informed, less wasteful, more efficient. Organizations of all kinds become wiser and more measured in their actions, able to treat each individual according to his or her just deserts.

But the examples above support a different picture. They remind us that personal-decision systems, left to themselves, rarely do better than promote the particular interests of their sponsors. Every such system bears the "fingerprints" of its creators—as when tenant-screening systems created by the real estate industry reward tenants' acquiescence and punish active efforts to resist poor rental conditions. Or when political campaign strategists rely on geofencing to facilitate electoral participation by those identified as supporters, while ignoring other voters. Cumulatively, these growing abilities of resourceful organizations to apply new decision-making powers change things profoundly, including changes not anticipated when the systems are set in motion. These include changes in what obligations are enforceable, whose interests dominate the treatments we receive, and our ability to influence that play of interests between ourselves and organizations.

Consider the changing character of our movements in "anonymous" public settings—as pedestrians on busy streets, spectators in a crowd, or travelers picking our way through train stations or air terminals. Many find this experience of anonymity refreshing and relaxing—our presence, though hardly a secret, goes unnoticed as we lose ourselves in the crowd. But as the inventors of geofencing and similar technologies have discovered, the facts of our whereabouts in a particular place at a particular time hold considerable latent value that may give an advantage to one party or another, if only such information can be captured and mined for the insights it affords. The fact that such information holds no protected status in American law offers an invitation to ingenious entrepreneurs and government planners to set their imaginations to work. Whose interests could they further, what profitable opportunities might they exploit, what costly problems would they be able to solve, if only all this raw material for decision-making were monetized, instead of being "wasted"?

With all this in mind, investors are perfecting systems to recognize and track members of once anonymous crowds—and to act on such information. Thus, many of London's affluent shopping streets are now furnished with intelligent trash cans that identify pedestrians by signals from their cell phones. The software in these devices can identify passing pedestrians by their cell phone signals and track their movements, according to an account in *Gizmodo*.[4] When these sensors spot a consumer known to be a high-value target, they can bombard that individual with invitations and inducements to visit the shops willing to pay the system to accommodate their sales pitches.

As usual, government agencies are pursuing similar capabilities for different (though perhaps complementary) purposes. Police and antiterrorist agencies are perfecting facial recognition software that can scan large crowds, picking out figures of interest and tracking them. The FBI is seeking to create a comprehensive national data bank of facial images—a project greatly abetted by the inclusion of computerized photos from driver's licenses.

In our mixed and sometimes contradictory reactions to accounts like this, we see the elements of dilemmas encountered throughout this book. On the one hand, a person's presence on the crowded streets of a city or other normally anonymous setting represents a classic instance of what nearly everyone would consider public information. So long as we do not seek to conceal or disguise ourselves, we leave open the possibility that anyone might notice that we were there. But most of us, most of the time, would find it at least awkward to discover that special note was being made of our presence at that specific time and place—perhaps more than just awkward. What interests, and what parties, are directing this special attention, we would wonder—and what purposes do they have in mind in collecting these data?

Traditional thinking about public-private boundaries doesn't help much with questions like these. While we value privacy, most of us also count ourselves as believers in freedom of expression—the familiar First Amendment right to communicate about nearly any topic, including one's fellow citizens. We would hardly want to live in a world where we were forbidden to disclose that we saw Mr. Jones on the King's Road, when he apparently believed himself to be "lost in the crowd." But technologies, and the sophisticated human systems developed to exploit them, change things. If someone develops a system that will identify every person walking on the King's Road, and with whom they are walking, the concept of "public information" may have to be rethought.

Imagine that a Silicon Valley entrepreneur or university computer scientist invents a technique for recording and archiving anyone's appearance in every public place where he or she goes. Suppose further that the inventor assembles a team to publish the contents of these archives on specific individuals. These made-to-order volumes of data would of course be highly personal—yet composed exclusively of what nearly everyone would consider public information. They would certainly hold vast interest for many parties—for example, employers wondering what their employees were doing during business hours; the IRS, concerned as to whether taxpayers' lifestyles were consistent with their declared incomes; antiterrorist investigators looking into "private" citizens' interests and associations; and inquiring spouses interested in matching the subject's accounts of his or her movements with those from the archive. These archival reports could presumably be offered for sale—first of all, perhaps, in an exclusive edition of one copy, at a premium price, to the subject of the data, on an ironclad promise that no further copies would be released.

Notwithstanding the awkward possibilities of such a scenario, those seeking to de-anonymize public spaces—and there are many—will point to many prospective benefits of their work. Personal-decision systems relying on the new technologies could locate missing persons, for example, or dangerous criminals, or suspected terrorists. And who knows—some consumers could be enticed into shops, electronically recruited from the street by retailers.

But what about those—the great majority, I suspect—who would prefer to be "lost in the crowd"? Many, at least, will react to these new potentials with distaste or worse. At a minimum, these new systems strike us as creepy—even if we are unable fully to articulate the reasons for such creepiness. Some of those reasons involve a visceral reaction against being observed when one has no option of observing the observers. Being lost in a crowd isn't quite the same when one cannot be sure one is indeed "lost." And our ignorance of the interests likely to guide such watching is hardly total. Systems like these, in the long run, not only secure our availability for commercial appeals from nearby businesses, but also make our whereabouts accessible to many other parties with strong interests in where we are and what we're doing. Do we really want to accord all interested parties unchecked access to the details of our private lives in this connection?

Many, I imagine, would accept a certain loss of privacy to enable law enforcement agencies to troll the faces of crowd members in order to spot actively dangerous felons. But what about efforts to apprehend those suspected of running red lights, or of neglecting alimony or child support obligations, or participating in banned demonstrations, or underreporting taxable income? Reasonable people will surely differ in their answers to questions like these. But perhaps we can agree that the prospect of surveillance that could reach out *at any moment* to enforce *any obligation* would change something fundamental in our concept of citizenship. That *something*—the room for passive resistance, evasion, or foot-dragging in complying with public obligations—may measure the difference between an open (if imperfect) social order and a drift toward authoritarianism.

We are right to worry about the rise of systems capable of doing these things. Actually, we should worry about them in a number of quite different ways. In the first place, powerful means of knowing about people and of acting on that knowledge trigger outrage for *not working as intended*—for failing on their own terms. They may incorporate inaccurate or outdated information in people's files, for example, thus multiplying the effects of a simple mistake. Elsewhere, personal-decision systems may—in data breaches, for example—disseminate images or details of people's lives that the subjects find dangerous, threatening, or deeply embarrassing. Even when no malice is involved, most of us do not want the fine details of our medical records disseminated to all and sundry, any more than we would want pictures taken of us in moments of extreme grief, joy, or embarrassment to be published. Like sophisticated medical procedures, personal-decision systems appear indispensable when they work well, and horrific when they go off the rails.

But we should worry still more about the prospects that these systems will one day work *much too well*. The result would be a world where power relations of all kinds penetrate more deeply into everyday life, with fewer checks and balances. There will always be those ready to insist that, if a certain measure of institutional enforcement power is good, more must always be better. If law-abiding behavior and civic responsibility are desirable, why not use every available means to these ends? Forgotten in appeals like these is that the *instrumentalities* of influence over people's lives are likely to long outlast the concerns and climates of their inception. Personal-decision systems are normally expensive and complicated to create. Once such costs are embodied in working institutions, the new capabilities are apt to remain available for uses that may have been no part of the original plan.

A classic case among American personal-decision systems is Social Security. This institution does extremely well at some very complex and costly tasks—above all, allocating unique numbers to millions of individuals, used to follow them and record their contributions to the system throughout their economically active years. When first proposed, the system attracted distrust and suspicion from labor unions and working people who feared that employers would use the numbers to track and discriminate against pro-union workers. Spokespeople for the new agency insisted that its capabilities would serve only internal needs of the system.[5] But today, both Social Security numbers and the institution's enforcement capabilities are used for a wide variety of non–Social Security purposes—for example, enforcing child-support obligations on absconding parents. Such examples remind us that once systems of this kind are created and the "sunk costs" of operating them are absorbed, their creators can lose all control over their subsequent operations.

Consider the classic twentieth-century manifestations of state power run amok—totalitarian regimes bent on targeting vulnerable populations. Most notorious among these were campaigns in Nazi Germany and Stalinist Russia to reach deeply into the ranks of the governed and pressure those designated as enemies of society. Perhaps most horrific of all were Hitler's waves of domestic repression that sent millions of Jews and other minorities to death camps. The Netherlands had the distinction, at the outbreak of World War II, of possessing particularly thorough population records, including details of each citizen's religious background. Once under Nazi control, the central population registry permitted the occupiers to pinpoint Jewish citizens, many of whom were then deported to their deaths.

Efforts of the Dutch resistance to sabotage these activities culminated in what must be a unique action in the annals of privacy contention: the April 1944 Royal Air Force bombing raid against the central registry of personal information, housed in the Kleykamp mansion in the Hague. "The attack . . . had deliberately been planned to take place on a working day. The chances were greater that the archive cupboards would be opened and that personal identity cards would be placed on the desks, which would enhance the chance for destruction of this shadow archive." In the words of RAF squadron leader Robert Bateson, "We bombed from an altitude of approximately 18 meters, so lower than the roof. A sentry was posted outside, but he threw his rifle away and ran. . . . All bombs landed on the target and the incendiaries functioned as intended. When number 5 and 6 had had their turn nothing very much remained of the whole building."[6]

Many observers are convinced that the strength of Europe's privacy codes today stems from historical memories of events like these—in an era when modern personal-data systems were mobilized to serve the most ruthless regimes.

Americans, by contrast, often seem to regard such deadly applications of state power as an exotic aberration, with little relevance for this country. Such attitudes ignore cycles of repression in American history, moments that recur with disturbing regularity. In the twentieth century, the era of the Palmer Raids immediately after World War I, brought mass arrests of alleged "subversives," suspensions of civil liberties, and a pervasive (though fortunately relatively short) climate of intolerance for dissent. The 1950s saw the McCarthy era, with its efforts to root out suspected communists and their alleged sympathizers from all corners of American life. The 1960s brought popular agitation for civil rights and for peace, countered by FBI campaigns against both these movements. Those efforts included illegal FBI burglaries of opposition groups' premises and even attempts to drive Dr. Martin Luther King Jr. to suicide. Can anyone doubt that these strictly political dynamics would have left far more destructive consequences, had today's sophisticated personal-decision systems been available to the authorities?

The Need for "Legal Bases"

"This is not bananas we are talking about," Spiros Simitis drily noted in a 1999 interview with a reporter for the *New York Times*.[7] One of Europe's first and most distinguished personal data protection commissioners, Simitis was comparing European Union debates over personal information with those on agricultural import policies. His remarks articulated what, by then, had become a consensus view—that personal information required treatment different from that given just any asset or resource. The potentially intolerable social costs when such systems go wrong mean that powerful strictures are indispensable for their guidance.

Thus, we need to reimagine personal-decision systems as belonging to a special category of technologies, activities, and social relationships that we rightly approach with deep ambivalence. This category includes, for example, the use of poisons and addictive drugs, access to particularly dangerous weaponry like rapid-fire rifles, transactions such as the sale or donation of human organs, and potentially problematic or exploitative social relationships like surrogate motherhood. All these disparate things set in motion powers and forces too sweeping to be left strictly to the discretion of those who create them. Though often valuable or even indispensable in their proper contexts, the costs of these activities' going wrong are so great as to require precisely targeted regulation.

From the beginning, European legislators have applied this logic to personal-data systems by requiring that every such system possess a *legal basis*—that is, an officially recognized, socially desirable function. This policy amounts to imposing a filter, limiting the permissible forms and purposes of such systems. Thus, my proposed Reform One:

> The United States must adopt requirements similar to Europe's, making legal bases indispensable for the operation of all personal-decision systems, private or governmental. Essential to any such a legal basis is the promise of the system to serve a broad range of public interests, and not simply to promote the interests of its original sponsors or any other single party.

It is essential that the owners of every personal decision system covering more than a minimum number of subjects go on record with their account of the larger social needs served by the system. This account should take the form of a succinct statement filed with their other declarations on the system.

Note that the requirement for this declaration, discussed in more detail in chapter 4, does not amount to a system of registration and inspection by government authorities. It becomes a focus of action only when outside parties—data subjects, activist groups, or responsible agencies like data-protection bodies—challenge either the aims or the execution of the statement. The ready availability of these declarations to public scrutiny would represent the first step toward creation of a profoundly new informational environment for all Americans—one in which the workings of personal-data systems are far more accessible and open to intervention than today.

In the European Union today, acceptable legal bases for most personal-data systems are easily stated. Under the General Data Protection Regulation (GDPR), situations in which bureaucratic processing of personal information is acceptable include those (1) where the recordkeeping exists to support a contractual relationship between an organization and an individual (bank accounts, periodical subscriptions, mail-order accounts with retailers, etc.); (2) where recordkeeping is required by law (in the United States, for example, where pharmacies are required to keep records of sales of certain drugs); or (3) with the consent of the subject (as in membership lists of political parties and other voluntary associations).

Oversimplifying just a bit, we can divide personal-decision systems into two broad categories. First are *interactive* systems, where we can assume that both the subject of data processing and the institution share some interest in the continued operation of the system. Examples include systems associated with credit cards or with accounts between landlords and tenants, or between tax authorities and taxpayers. In these and many other cases, the parties recognize some necessity of maintaining an ongoing relationship, and associated records of payments, dates, disputes, and settlements—all of which amount to personal information.

But another category of recordkeeping, apparently growing rapidly in the United Sates over recent decades, consists of what I term *unilateral* systems. These are the work of organizations that compile and manage personal data on subjects with whom they have no legal or contractual links. These include major databases of many data brokers—like Acxiom, Exact Data, Paramount Direct Marketing—and countless other businesses, large and small, that specialize in "background reports" of various sorts for purposes of evaluating job candidates, checking on prospective business associates, or even screening prospective dates. These operate legally in this country without the consent, and often without the knowledge, of those reported on. Following the reform proposed here, such systems would have to cease operations or radically change their business plans.

This would amount to a sweeping interdiction indeed, potentially resulting in the shuttering of businesses that have flourished and prospered during the expansive years of the internet revolution. These would include the landlords' information systems that blocked Margot Miller from escaping from homelessness, which function essentially as blacklists for nearly everyone included in their files. This severe proscription reflects my judgment that the asymmetry of resources between the organizations involved and the subjects of their attention is simply too profound to be acceptable. Under these circumstances, we need to fashion much more effective institutions for protecting the interests of grassroots targets of professional surveillance. Discussion of some such possibilities follows later in this work.

Just to make judgments on these matters still more subtle, many organizations effectively offer services based on personal record-keeping to take place in the future. The XYZ company may propose to book you a table at your favorite restaurant; or teach you a foreign language in online lessons; or even set you up with dates with scientifically selected prospects—but only if you agree to allow this company (along with its "partners," "affiliates," and all and sundry others) to share the personal data to be extracted from you. In short, this is the familiar offer of "free" services, purchased in the coin of open-ended plunder of one's privacy.

My position here is that there must always be legal alternatives to such Faustian bargains. Hence the need for Reform Two:

> Renunciation of one's privacy rights must never be a condition for access to products or services, or information about them. Organizations may make the familiar offers of "free" access, in exchange for personal data on the users—provided that the data and its intended uses are correctly specified. But those who make such offers must offer the identical services to all customers for a cash fee, without the capture of personal data, at their "shadow price"—that is, a fee no greater than the market price of providing the service in question. Shadow pricing is a technique economists use to estimate what the price of a given good or service would be, if there indeed were a market for such a good or service. Thus, planners in a country where strawberries were unobtainable in winter might calculate a "shadow price" for strawberries, based on what consumers are known to be willing to pay for other warm-weather fruits that are sold at that time of year. Here, as elsewhere in these proposed reforms, the aim is to preserve for everyone the option of carrying out transactions without loss of privacy.

None of this is to suggest that these first two reforms should imply that all demands for personal information by organizations need be met with compliance. One of the most troubling reflexes that all of us develop in the privacy-challenged world we inhabit is that of obediently filling out forms or otherwise providing personal data that organizations do not really need. These routines are so deeply wired that we often neglect to ask, "Is this necessary?" Often, it turns out, the relationship or transaction in question could go forward perfectly well without collection of any personal information at all. In the words of UCLA law professor Jerry Kang, personal information may be "functionally unnecessary" for what we are seeking—booking a trip by train, for example, or purchasing over-the-counter medications, or logging on to a website. If the institution carrying out the personal-data processing can present no rationale as to why the information is indispensable for the transaction in question, no one should have to provide it.[8] Thus, I propose Reform Three:

> No organization may require personal information from anyone to complete any transaction or provide any service, unless the organization can establish the "functional necessity" of that information in that context. Accounts of such functional necessity must be provided, along with other data on every personal-decision system involved (as detailed in chapter 3).

As many privacy-watchers have held, this reform should represent a universal defense against gratuitous erosion of privacy. Every member of the public should be prepared, upon adoption of this recommendation, to query those who seek personal information for the organization's justification of its need. Spokespeople for the organizations involved should have their speeches prepared in advance on this point—or be ready to drop their request for the data in question.

No Legal Basis, No Personal Data Use

Operation of personal-decision systems without legal bases must simply be proscribed. Activities like those of Acxiom and other companies that appropriate personal data from disparate sources to sell for advertising or marketing purposes should become impossible without explicit consent from the subjects of the data. One felicitous result would be greatly to reduce the blizzards of unsolicited appeals that crowd Americans' electronic and physical mailboxes. Another would be to block reuse of one's personal information, supplied to companies like Google and Facebook, for the targeting of future appeals—both those going directly to their customers and those used to direct ads for their advertisers. When subjects find themselves targeted by unwanted appeals via any personal-decision system, they must have the right to know the source of all data on themselves included in the system, and to withdraw from the system altogether at their own discretion (as proposed in chapter 5).

Lacking a categorical right to privacy applying across different contexts, the United States has de facto embraced the other extreme. The result is a legal order where, as one European observer has commented, "all forms of the processing of personal data [are] . . . possible unless they are forbidden in concrete cases." So long as business activities are not covered by any of this country's "sectoral" laws—governing video rentals, for example, or health care delivery, or certain financial records—one is free to compile and market virtually any form of personal data. The reforms proposed above would obviously run in a collision course with these principles—and would accordingly result in the cessation of an array of long-established practices.

But one can imagine accommodations that might establish better protections for privacy interests, while still allowing the essentials of some systems to be preserved. If these organizations could find ways to win the express *consent* of their data's subjects to their operations, they would fulfill a basic condition for their legal operation. To attract such consent, they would no doubt have to offer something in return, presumably some additional service or product that would make an ongoing relationship attractive to subjects of their recordkeeping.

Such compromise—soliciting consent from some subjects, and thus establishing a legal basis for their operations—could work where the two parties start with at least some shared interests. Establishing the consent of data subjects would bring the watchers and the watched into closer contact—raising the probability that the latter may become aware of the tracking and understand their ability to contest or even block it. But it is hard to see solutions like this working in the many cases where unilateral systems function essentially as *blacklists*—where inclusion virtually never means anything but serious bad news for those covered. These include landlords' systems for identifying undesirable tenants, systems devoted to reporting criminal records and other disreputable behavior, compilations of bad debts (accounts overdue sold for collection), and the like. These systems are particularly dangerous for privacy and equity interests, in that their owners are motivated to produce strictly negative information for use by their buyers. Accordingly, those who pay for reports have little interest in inputs from those reported on. The asymmetry in these positions makes them accidents waiting to happen, in privacy terms. In their pure form, they should never enjoy legal status.

The response from the sponsors of these systems is predictable. They will claim that their surveillance activities represent nothing more than elementary self-defense—against irresponsible tenants, felonious would-be employees, hotel guests who stay but don't pay, dangerous would-be dates, and the like. Pushing back against such efforts of self-protection, they will insist, can only trigger outrage from righteous business owners seeking honesty in a dishonest world. There is no reason to discount such complaints across the board.

The only solution in such cases, it appears, is administration of such functions through some public agency, such as a data protection commission or (at the state level) a public utilities commission. Such an agency would be charged to develop its own standards for admissible and inadmissible entries, rights to challenge the contents of one's file, and a body of precedents for resolution of recurrent conflicts. Anything short of this active institution building risks leaving the fingerprints of highly self-interested industries on a host of relationships in which institutions now hold the upper hand and clients now have no advocate for their position.

Government Systems

Many observers have contended that Americans are more suspicious of government surveillance than of that by private-sector interests. The fact that government agencies can compel both creation of and access to personal data—and the sheer extent of resources available to government institutions for doing these things—fuels fears of abuse among Americans both on the right and the left. These anxieties grew dramatically in the Watergate period, culminating in the impeachment of President Richard Nixon and his resignation in 1973. Public mistrust of government was at its apex. Much popular uneasiness focused on data gathering by the FBI. Reinforcing such anxieties were the allegations long circulating in Washington of "secret files" of scandalous information on American politicians developed by FBI director J. Edgar Hoover—and used by him to influence American politics through more or less veiled threats of disclosure.

Those rumors might have been dismissed as paranoia—until the discovery of just such an archive of scandalous reports at FBI headquarters following Hoover's death in 1972. In the words of federal judge Laurence Silberman, then serving as acting attorney general:

> I was shocked ... when on January 19, 1975 ... I read a front-page story in the [Washington] Post confirming the existence of the files. ... I read virtually all of these files. ... It was the single worst experience of my long government service. Hoover had, indeed, tasked his SAC's [special agents in charge] with reporting privately to him any bits of dirt on political figures such as Martin Luther King, or their families. It is also true that Hoover sometimes used that information for subtle blackmail to ensure his and the Bureau's power.

> I intend to take to my grave nasty bits of information on various political figures—some still active.[9]

Just weeks before the appearance of the *Washington Post* story cited by Silberman, Congress had passed the Privacy Act of 1974. This was a direct response to Watergate-era alarm over government surveillance over Americans' lives. But federal government surveillance was by no means the only source of Americans' rising privacy unease. The keeping of personal records of all sorts was increasingly seen as a threat to freedom of thought and action. As Sarah Igo, in her authoritative history of privacy in America, characterizes America's public mood,

> collections of data were not just intrusive or irritating features of living in the modern age; rather they *did* things to people and maybe even changed the nature of personhood. Data banks had a way of ricocheting back on the person, inviting questions about what a "subject of data" really was. "As the public becomes increasingly aware of the information orientation of modern life," wrote [Harvard Law Professor Arthur R.] Miller, "it is understandable that people may begin to doubt whether they have any meaningful existence or identity apart from their profile store in the electronic catacombs of a 'master' computer."[10]

A pervasive cultural malaise like this, coinciding with a political crisis embodying major themes of government snooping on "private" citizens, provided a combustible mixture. A key response from America's officialdom was the Privacy Act of 1974.

Unlike nearly all other American privacy statutes, the Privacy Act follows what today we think of as the European model: setting down a series of general principles and applying them across a variety of settings. The Privacy Act does not govern all personal-information systems, but it applies to many systems maintained by federal government agencies, and those of certain federal contractors. Above all, it sets down procedures for subjects of federally maintained records to inform themselves of the contents of their files, and to challenge entries that they find unsatisfactory. These provisions take their inspiration from the much-heralded Fair Information Practices detailed in the preface to this book.

Consistent with the Fair Information Practices are the Act's reporting requirements. They require publication in the *Federal Register* of details on each personal-data system held by every federal agency—excepting those involved in investigatory activities of the CIA, law enforcement, and, in fact, a number of other federal bodies such as the Department of Homeland Security.[11] The boundaries of this important exclusion remain elastic and subject to change, as agencies seek to exempt portions of their operations from coverage. Nevertheless, the reporting requirements make it possible to gain some overview of federal personal-decision systems, which number roughly ten thousand at the time of this writing. They are a remarkably diverse lot, ranging from those whose existence is more or less assumed—like the taxpayer records maintained by the Internal Revenue Service—to the relatively obscure record systems developed by USA.gov, an office devoted to helping Americans locate unclaimed funds owed them by government agencies.

Prominent or obscure, all these systems share the potential for major impacts on the lives of those depicted in them. That is why Reform One holds that the requirement for legal bases should apply to all government systems, at all levels of government, as well as to those in the private sector. The permissible legal bases for government systems should closely parallel those for private-sector systems. Personal-decision systems maintained by government agencies should be acceptable (1) when demonstrably necessary to carry out the statutory mandate of the department—as with data systems on drivers for state motor vehicle departments; (2) when part of a quasi-contractual relationship between the individual and the agency, as when individuals request to be informed of public hearings on a particular form of federal activity; or (3) with the consent of the individual.

Note what these legal bases do *not* include. They grant no place for systems created solely at the discretion of a government agency, without either fulfilling a legislative mandate or involving willing cooperation or assent from data subjects. This requirement would exclude, for example, data systems created to track journalists or members of the public critical of the agency. Most Americans would probably agree that such recordkeeping represents gross misuse of government resources. Such unilateral systems should be proscribed for lack of a legal basis.

Some readers will react with skepticism to any proposal to subject government record systems to requirements for legal bases. Won't any and all government departments be quick to insist that *all* their record systems, including personal-information systems, are created only for the most high-minded of purposes? And if that fails, won't these agencies simply turn to their allies in Congress for blanket approval of any and all personal-data systems maintained by the agencies as indispensable to their missions? Here as elsewhere, getting highly self-interested actors to police themselves doesn't sound like a winning strategy.

Admittedly, it won't always work. The record to date demonstrates the expertise of federal agencies—the FBI in particular—at subverting attempts to rein in controls on personal-data-keeping that they find inconvenient. Perhaps J. Edgar Hoover, had he sought to do so, could have gained congressional approval for the Bureau's notorious "black bag jobs" from the 1940s to the '60s, in which agents secretly broke into private premises, without court orders or search warrants. But having such operations legally sanctioned, simply because they were carried out by any specific agency, regardless of statutory authority, would represent the worst of all possible worlds from the standpoint of privacy.

Legislation requiring legal bases for government as well as private-sector personal-decision systems would be worth having, even if Congress and the courts initially ratified such bases with excessive complaisance. Such unsatisfactory outcomes would at least place all parties on notice that every such system requires some higher justification than the convenience of the agency. Assuming the creation of some form of national authority for privacy protection, having such legislation on the books could open the way for that authority to challenge the existence of personal-decision systems that appeared to serve no higher social interest whatsoever.

Here, some agencies will rejoin in much the same terms as private-sector organizations: how are they to protect the legitimate interests of the data-keepers from those bent on harm—for example, people who have attempted to defraud the agency? The answer in this case is to require special and explicit approval from Congress. That justification should be entered, along with the entire list of personal-decision systems maintained by every agency and submitted for congressional action. The need to "go public" in this way should be salutary for all concerned.

These steps seek to promote a key aim of this work—to *politicize* privacy, in the best sense of that term. But privacy-watchers can attest that prevailing pressures from government agencies are pushing in the opposite direction—toward exclusion of existing systems from the reach of privacy protections already in place.

The Privacy Act explicitly places investigatory files of the CIA and law enforcement agencies beyond the reach of its strictures, along with record systems in certain other agencies deemed to be maintained for law-enforcement purposes. Moreover, in a strikingly self-defeating twist, the Act places the power to make these exemptions in the hands of agency heads. The FBI has pushed back against the spirit of the Act quite successfully, by invoking this power. In 2017, it succeeded in placing the entire contents of its massive Next Generation Identification (NGI) system off-limits to inquiries from citizens seeking records on themselves contained in that system. The system holds such records—including fingerprints, photos, and biometric identification—on more than a hundred million people. Many of these are convicted criminals. But many others are simply people who have submitted their fingerprints or other data in connection with background checks needed, say, for employment. It does not take a fortune-teller to foresee the ultimate aim of the FBI in this maneuver, given the ever-growing array of sources of identifying information available to feed the system. Avowed or not, that aim is surely to ensure the ability of the Bureau to identify and track the great majority of the U.S. population, regardless of the circumstances that brought their data under control of the NGI system.

Nearly everyone can agree that some restrictions on access to investigative files are indispensable. Law enforcement and counterintelligence agents would be in an impossible position, if the targets of their investigations could determine, at any given moment, whether their activities were being scrutinized. The question is how far the veil of concealment should extend. What about investigations that conclude without any enforcement action—or, indeed, in which the investigators concluded that there was no basis to investigate in the first place? Many subpoenas and court orders require that the targets of investigation be alerted to the information being sought—at least after the conclusion of the investigation, and sometimes from the very onset. This important step at least makes it possible for those targeted to react to the (probably unwanted) attention that they have received—and to react, if they feel that the grounds for the investigation were defective or capricious.

Still more to the point: many of those whose data are now incorporated in one or another element of the NGI system are no criminals—past, present, or future. Yet the FBI has exempted itself from requirements to grant them access to their own NGI records, "to ensure that ongoing investigations are not compromised by people learning that they are the subject of probes."[12] Surely this development points in the direction of exactly the state of permanent government surveillance of all citizens that the original supporters of the Privacy Act were determined to combat. It is the very sort of investigatory overreach that ought to be blocked under a requirement for explicit legal bases for law enforcement surveillance.

Notes

1. Lisa Sanders, "What Caused Poison-Ivy-Like Blisters All Over This Man's Hands?," *New York Times*, Dec. 15, 2016.

2. Kim Barker and Jessica Silver-Greenberg, "On Tenant Blacklist, Errors and Renters with Little Recourse," *New York Times*, Aug. 16, 2016.

3. Ibid.

4. Kelsey Campbell-Dollaghan, "Brave New Garbage: London's Trash Cans Track You Using Your Smartphone," *Gizmodo*, Aug. 9, 2013.

5. Carolyn Puckett, "The Story of the Social Security Number," *Social Security Bulletin* 69, no. 2 (July 2009), https://www.ssa.gov/policy/docs/ssb/v69n2/v69n2p55.html.

6. Pieter Schlebaum, "Airraid on Kleykamp," trans. Fred Bolle, *Traces of War*, https://www.tracesofwar.com/articles/2441/Airraid-on-Kleykamp.htm.

7. Edmund Andrews, "Europe and U.S. Are Still at Odds over Privacy," *New York Times*, May 27, 1999.

8. Jerry Kang, "Information Privacy in Cyberspace Transactions," *Stanford Law Review* 50, no. 4 (Apr. 1998): 1193–1294, 1249.

9. Laurence Silberman, "First Circuit Judicial Conference" (speech, Newport, Rhode Island, June 19–21, 2005).

10. Sarah Igo, *The Known Citizen: A History of Privacy in Modern America* (Cambridge, MA: Harvard University Press, 2018), 243–44.

11. "Privacy Act of 1974: Implementation of Exemptions; U.S. Department of Homeland Security/U.S. Immigration and Custom Enforcement-018 Analytical Records System of Records," *Federal Register,* Nov. 8, 2021.

12. Ellen Nakashima, "FBI Wants to Exempt Its Huge Fingerprint and Photo Database from Privacy Protection," *Washington Post,* June 1, 2016.

3   *Require Consent for Disclosure*

Early voices on the risks posed by large-scale personal-data systems proposed some straightforward countermeasures. The influential 1973 U.S. government report *Records, Computers, and the Rights of Citizens* put matters simply but forcefully: "There must be a way for an individual to prevent information about himself [*sic*] obtained for one purpose from being used or made available for other purposes without his consent."[1] The essence of this idea appears in a number of key early statements—including those from the Organisation for Economic Co-operation and Development (1980), the Australian Privacy Charter (1994), and the Canadian Standards Association (1996).[2] Certainly, the underlying principle resonates with our everyday conduct. When we disclose personal information that is not common knowledge, we often assume that the recipient will not act on it without our permission to do so. If we learn that this assumption has been violated, we are likely to reconsider the relationship.

Reading between the lines of *Records, Computers, and the Rights of Citizens,* one senses the authors reacting to a certain nightmare vision of a world to emerge, should their advice go unheeded. In such a world, personal information made available to *any* organization for *any* purpose would, in turn, flow seamlessly to *any other* interested party for *any other* purpose. In the recommendation quoted above, the authors appear to be appealing for systematic *compartmentalization* of personal data held by organizations, to countervail against such outcomes.

If so, history has been cruel to their vision. The trajectory of change since they wrote their report has run in quite the opposite direction—toward frictionless flow of personal information among organizations, and away from compartmentalization. In the United States, above all, both private-sector and government organizations have flourished by finding ever-more-innovative ways of obtaining and using personal data that originated in one setting for strategic purposes elsewhere. So dependent are American organizations on such cross-contextual exchange that efforts to require the subject's consent for release of data are apt to be cast as attacks on Innovation—the first cousin of Technology itself, apparently. In the words of one industry enthusiast, Heidi Messer:

> We live in a networked world. The internet is built for sharing things at little to no cost. We forward our emails, capture photos on cellphones and tweet opinions, all activities that leave a trail of data that can be collected without our knowledge. Privacy—the right to be free from unwanted intrusion—no longer exists in an absolute sense. . . .
>
> The biggest companies—led by Facebook, Amazon, Netflix, and Google in the United States, and Baidu, Alibaba, and Tencent in China—are data networks, aggregating information to provide valuable services underwritten by advertising, e-commerce or user subscriptions. If we constrict their fuel—data—we may hurt not only the quality, cost and speed of their services, but also the drivers of growth for the world's economy.
>
> Innovation will . . . suffer.[3]

The vision first articulated in *Records, Computers, and the Rights of Citizens* does not deserve this unceremonious neglect. The authors of the 1973 report were right to foresee the disastrous consequences for privacy of yielding to pressures for unrestricted free flow of personal information among interested parties. Yet the very logic of many personal-data systems requires some form of sharing or cross-checking.

The question, then, is *how much* sharing of personal information are we prepared to sanction, and in *what forms*—and where do we properly hold institutions responsible for keeping filed information private and compartmentalized?

If we are to take privacy seriously, the new default condition must be Reform Four:

> No data held in any personal-decision system may be used, shared, or disclosed to any other party, except as required by law or with express consent from the subject.

Note that this reform presumes a bright line between a personal-decision system as described in chapter 1 and countless other forms and collections of personal data. A congeries of bits and pieces of personal information becomes a *personal-decision system* when (1) the data are held in readily accessible form in an interconnected system, subject to routines of retrieval; and (2) that retrieval serves to support decision-making to shape action on the individuals concerned. Thus, Richard Nixon's alleged "enemies list" of political nemeses would not qualify (since it was apparently not committed to writing, nor did it involve retrieval of data from identifiable databases). But the computerized listings of persons not to be admitted to the United States, as used at the nation's borders by Immigration and Naturalization Service staff, definitely would.

Nothing in this reform would curtail anyone's efforts to speculate about personal information that might or might not exist, whether held in a personal-decision system or not. We would not want to live in a world where journalists—or anyone else—were blocked from efforts to suggest that an elected official, for example, was living far beyond his or her means, given the official's publicly declared sources of income. Nor would we want strictures that prevent a historian writing a critical biography of a living figure from suggesting that the latter's public appearances indicate an illness that has not been publicly acknowledged. But *direct release* of personal data from a personal-decision system held in the database of an accountant or a physician would require consent from the subject.

"Personal information" should be understood here in the broad sense. It includes, for example, information derived from "cookies" placed on computers by those bent on tracking the internet activities of users. It also includes the accumulation of what Shoshana Zuboff brackets as "behavioral surplus" derived from analysis of users' interactions with search engines, social media, and the like—insights into intimate tastes, assumptions, and sensibilities that might afford levers for manipulating the subject's actions.[4] Thus, models of consumer behavior like those derived from Google's vast troves of internally generated information, along with personal data purchased on the market, would be proscribed by this reform.

No such information should be subject to use or disclosure by the organization holding it, except as required by law or with the subject's consent. Relations between subjects of such data and organizations holding the data would thus come to resemble *fiduciary* relationships, as Daniel Solove has suggested.[5]

Private-Sector Disclosure

The decades since the publication of *Records, Computers, and the Rights of Citizens* have hurtled us light years away from any such compartmentalization. Like it or not, Americans today widely assume that unauthorized collection and sharing of their personal information is all but universal. A few forms of such unauthorized appropriation are conspicuous to nearly everyone—for example, the use of data gleaned from our internet activities to target ads served up to us by Google, Facebook, and other social media connections. But few Americans have dug as deeply, in trying to find out who is appropriating and using the information about themselves that flows through the internet, as Brian X. Chen. A veteran *New York Times* columnist, Chen covers information technology. In a particularly telling article, he described his exploration of data on himself compiled through his Facebook page.

With the briefest of efforts, Chen found that some five hundred advertisers were holding data on him—including a number of businesses he had never heard of, like a motorcycle parts store and an electronica band. Many of these had his contact information, including his email address. One social network had his entire "phone book," including the number needed to activate his apartment buzzer. The social network "even kept a permanent record of the roughly 100 people I had deleted from my friends list over the last 14 years, including my exes."[6] Chen was unhappy to find that information he considered particularly sensitive, like the names of people he had unfriended, remained permanently in the Facebook record, even if no longer visible to visitors to his page. And particularly alarming was the fact that personal data from his Facebook account had found its way into the databases of those hundreds of retailers—most of which he had never interacted with.

Under the principles put forward in chapter 1, many such essentially *unilateral* collections of data would be illegal for lack of a legal basis. But given today's mind-set, Americans are likely to consider such one-sided feats of data appropriation the inevitable effects of Technology—and hence normal business practice for the giants of our information industry. Less well understood is the pervasiveness of such practices among lower-profile companies, most of them unknown to us by name or function.

Think of the frustrations of Margot Miller in Harlem, as she sought to escape homelessness (see chapter 2). How could she have known that New York's housing courts were selling data from their records to companies eager to spot tenants who had been in court with their landlords? The many organizations who make it their business to furnish such fateful information often prefer to keep a low profile, if only to shield themselves from repercussions from aggrieved members of the public. Under Reform Four, those data would not be the landlords' (or the courts') to sell—without consent from those depicted in the files.

True, some forms of personal information in this country already enjoy certain protections under sectoral privacy laws—video rental records, for example, or certain health or financial transactions. Statutes, many dating from the 1970s and '80s, forbid release of such information at the sole discretion of the holder of the data—though these same sectoral laws also specify many situations where holders of such data files have virtually no choice but to open them to law enforcement or other state agencies. HIPAA, for example, the Health Insurance Portability and Accountability Act, while largely protecting the privacy of medical records, does not require patients' permission to release their personal information to law enforcement agencies.

Beyond these islands of ambivalent protection, sheer *access* to personal data is understood as tantamount, in many settings, to the right to sell, exchange, or otherwise transfer it. U.S. law protects companies' prerogatives to do as they wish with the personal information they acquire—opening the way for telecom companies to sell their subscribers' calling logs, for example, to retailers and advertisers for marketing purposes. The courts have essentially held that consumers have already relinquished control over "their" data by confiding it to the carriers. This principle marks a profound divergence from the logic of Europe's privacy regulation, which blocks most release of personal data from files created for one purpose to new destinations, without the subject's consent.

For some readers, the spread of personal information in this fashion spurs little concern. Much of the data, they will insist, is so banal or routine as to be inconsequential. And what's more, the unhindered flow of such data results in all sorts of "free" goods, services, opportunities, and stimuli tailored to the interests and needs of the user—rewards that he or she may not even have been aware of, before encountering them on the internet.

But privacy-minded analysts will insist that one can never foretell what uses or practices will govern the repercussions of collecting any form of personal information over the passage of time. After all, new ways of combining and interpreting even the most ordinary and public of personal data may yield insights that leave the individual at some unexpected disadvantage. Thus, if an artful analyst in the Department of Homeland Security were to establish a strong statistical association between Basque ancestry, residence in a zip code ending in 09, preference for a certain brand of toothpaste, and terrorist sympathies, any rational individual sharing those characteristics would be well advised to take evasive action immediately. No one can foretell what interests will guide the use of such ever-emerging associations.

. . .

It would be reassuring if we could point to some bright line between personal information whose dissemination no one could object to and data so sensitive that disclosure poses inevitable risks to the subject. But this is a fond hope. The very facts about ourselves that we regard as most strictly "personal" often hold the most intense interest for other parties. Think of information on bodily states, symptoms of disease, and treatments we seek to deal with them.

Most people do not expect to disclose such data, other than to intimate acquaintances and caregivers. Yet, at the same time, sellers of drugs and health-related services are desperate to obtain personal information that might identify would-be buyers. Innovative companies have thus perfected methods for "stripping"—that is, copying—patient data from electronic orders routinely sent from pharmacies to suppliers. The stolen data may used to to target appeals to physicians, and sometimes directly to patients, to adopt alternate choices of drugs.

None of us likes the thought of anonymous but highly interested parties monitoring what medications we're taking—and, implicitly, the illnesses making such medications necessary. Sometimes, aggressive uses of prescription data can trigger considerably more serious consequences than unwanted advertising appeals. Recall the experience of the Louisiana couple discussed in the introduction, who found that their medical insurance rates had been raised, apparently because the wife had once taken a drug often prescribed for high blood pressure. Many companies profit from the sale and trade of data obtained from pharmacists on people's prescription histories, or indeed from information people confide to websites on their current medical conditions. Such information is much sought-after by otherwise unidentified "partners" or "marketing affiliates" of the companies collecting it.

Sometimes these mysterious "partners" operate under contractual agreements with the companies from which personal data are "scraped." But elsewhere corporate users of personal data raid one another's systems. One target of such raids has been PatientsLikeMe.com, a site that offers forums for those suffering from various diseases to discuss their symptoms. What most users of the site apparently do not realize is that the personal information they disclose there is subject to capture, resale, and reuse in countless personal-decision systems whose identities are likely unknown to the subjects.

Since health care delivery represents such a major slice of the national economy, the ability to direct the flow of these expenditures matters enormously. Actionable information as to who is being treated for which symptoms, if delivered promptly and in the proper form, is tantamount to a license to print money. In the scrum to provide such data to paying customers, data brokers troll for patients' email addresses, personal websites including résumés, social network postings, and countless other sources in the personal-data-rich environment we now inhabit. According to a *Wall Street Journal* report by Julia Angwin and Steve Stecklow, companies compete intensely with one another to capture, or "scrape," personal data en masse from social networking sites and various other online meeting-places where Americans share personal information on their health. These companies then resell the data to interested parties that produce individual "background reports" for purchase over the internet. Resale of "publicly available information" included in these downstream transactions apparently poses no legal issues for the seller.[7]

Practices like these run in a collision course against the letter and spirit of the principle upheld by the authors of *Records, Computers, and the Rights of Citizens* in 1973. Nielsen, and countless other companies, take for granted their right to personal information whose value arises precisely from the fact that those contributing to it are oblivious to its appropriation. Unfortunately, this depressing development dramatizes the sweep of historical change in the management of personal information since 1973.

PatientsLikeMe.com is by no means unique. Because Americans' consumption of medical care triggers such vast expenditures, wily innovators have converged on consumers' quest for reliable health information like birds of prey. Thus, it is difficult to rely on the internet for information on health issues *without* having one's personal information captured by those with something to sell. In a rigorous review of these practices, computer scientist Tim Libert of the University of Pennsylvania surveyed more than eighty thousand web pages devoted to such purposes. He concluded that appropriation of personal data from such sources represented the rule rather than the exception for many companies in this industry. One company, Medbase 200, was reported as using proprietary models to generate and sell lists with classifications such as "rape victims," "domestic abuse victims," and "HIV/AIDS patients."[8]

We can assume that most of those being tracked in this way have no clue how easy it is to capture the often quite sensitive information that visitors inadvertently provide to these sites. In some cases, technological countermeasures may help thwart such efforts—for example, by suppressing "cookies" left on computers to track the websites visited by their users. But American industry continues to generate unobtrusive workarounds to defeat such privacy-seeking steps.

In one particularly notorious case, the Federal Trade Commission (FTC) in 2012 announced a settlement with the giant online advertising company Epic Marketplace, after challenging the company's "history sniffing" technologies for tracking consumers' online queries. According to federal government sources, Epic had promised to collect only information about customers' visits to its own websites, but instead used these new tools to collect patient data from other sites as well. The FTC contended that Epic had used these captured data to group lists of patients into categories like "Incontinence," "Arthritis," "Memory Impairment," and "Pregnancy-Fertility Getting Pregnant." Epic then used these databases to direct ads to patients, according to the categories in which they were listed.[9]

"History sniffing"—a marvelously evocative name for this rapaciously privacy-destroying practice! It captures the sly tactics eroding privacy in countless realms of twenty-first-century life. What Epic Marketplace had perfected was a highly effective means for linking details of Americans' past histories with their propensities for future consumption. Highly effective, and mostly inaccessible to the individuals who were ultimately bombarded by the targeted entreaties of Epic Marketplace. What about a consumer who is definitely not expecting, or expected by others, to become pregnant—but who nevertheless finds herself carpet-bombed by ads suggesting that she is? In at least one such case, a woman was surprised to be congratulated by an array of ads on becoming pregnant—only to miscarry a short time after getting the news. There have been reports of women receiving birthday greetings for a never-born (but badly wanted) child for years after its non-birth. The business of America may be business, as Calvin Coolidge is alleged to have insisted. But here, surely, the marketers have lost all restraint. As more and more contexts of everyday life are computerized, leaving revealing informational residues, the scope for "sniffing" in our pasts grows ever greater. And as the payoffs for successful prediction of the actions and interests of consumers grow, temptations for such linkage grow commensurately. Only strong legal codes can block such trends.

Indeed, such practices feed upon one another. The more of these novel and unobtrusive sources of data there are, the more ingenious exchanges among data systems there can be. Consider the remarkable innovations since the 1990s in the use of consumer credit scores. By that decade, the American credit reporting industry had succeeded in reducing consumers' entire credit histories to three-digit scores, encapsulating each consumer's desirability as a credit customer. Credit grantors—banks, credit card companies, mortgage grantors—apply their chosen algorithms to each applicant's score to determine instantly whether to grant any credit application, and on what terms.

Having perfected this sophisticated scoring system, the credit reporting industry then cultivated a new source of revenue in the insurance industry: persuading insurance companies that credit scores could predict the likelihood that insurance buyers would ultimately file claims. Thus, a stunning business coup: spotting a completely new market for a product that credit reporters already had "on the shelf." The insurance industry responded by purchasing credit reports on insurance applicants and adjusting the rates for coverage to reflect the risk presumably identified by applicants' credit scores. The result is that low-income applicants (whose credit scores are predictably low) face sharply higher insurance costs than higher-income applicants. Where not outlawed by state legislators or regulators, this new form of disclosure of personal information is said to be highly profitable for both industries. It is likely that most insurance buyers affected by the practice never understand the role that information on their credit plays in the cost of their insurance.

These practices pose serious ethical and political issues—issues associated with any reliance on "big data" correlations to determine the treatment given to specific individuals. Consider the case of applicants for auto insurance. Many Americans would probably consider it reasonable to take some consideration of past driving history in setting rates for future coverage. A driving record marked by many citations and collisions, one might reason, could well portend more of the same. But using credit scores to set insurance costs is different. Most of us would probably have similar problems with the use of eye color or skin color—or, for that matter, height or bone density—to set insurance rates, simply because one has little or no control over these variables. Bad driving might be correlated with total numbers of citations and accidents, and insurance providers will be quick to charge more to applicants with such records—though traffic conditions and sheer density of auto registrations within a given neighborhood or district might also account for the observed associations. But no cause-effect relationship holds for genetically stamped characteristics like race or height, so the use of such information in setting insurance rates is suspect. The same holds for applicants' credit scores. Taking privacy seriously requires that they, too, remain off-limits for use in determinations with no logical connection to credit. Yet, at the time of this writing, only a few states—including California, Hawaii, Massachusetts, and Maryland—limit use of applicants' credit scores or credit histories in setting insurance charges.

Another concealed and unconsented form of disclosure that would surprise most internet users involves the use of VPNs, or virtual private networks. These are the systems that allow individual computer users to connect with networks maintained by organizations that may be physically remote from them—their workplaces, for example. Some organizations provide their members with such connections at the organization's expense, to facilitate employees' work at home. Elsewhere, private parties subscribe to commercial VPN providers in the hope, among other things, of safeguarding the privacy of their communications.

The latter expectations are often illusory. As in other industries, providers of VPN services pay the closest attention to which of the personal data passing through their hands can be "monetized." When a would-be innovator spots a crucial connection between specific personal data disclosed on the network and something that needs to be bought or sold, fortune beckons. Like the drug companies that "strip" patients' prescription data en route from the pharmacist to the wholesale supplier, many VPN providers are vigilant to spot and capture marketable personal data. Researchers who analyzed the business strategies of a large sample of VPN providers concluded that some 40 percent of their sample were engaged in privacy-unfriendly practices that were surely inconsistent with the trust placed in them by their enrollees. Some of the VPNs injected malware or malvertising into the clients' data streams; nearly 20 percent did not even encrypt user traffic.[10]

But some sort of black belt for chutzpah in stealthy, uncompartmentalized use of personal data must go to Unroll.me, a startup dating from 2011. Ostensibly, the site was dedicated to helping internet users unsubscribe from electronic mailing lists. To deliver this service, Unroll.me would gain access to subscribers' electronic address books. It then used this access to identify patterns in users' buying practices, which it then made available—for a price—to parties offering competing products and services. This practice came into sharp relief in April 2017, when it was revealed that the ride-hailing service Uber had used Unroll.me to provide data on its users' use of its competitor Lyft.

If users of Unroll.me had penetrated the fine print of the company's privacy notice, they could have read that "we may collect, use, transfer, sell and disclose nonpersonal information for any purpose" and that the data can be used "to build anonymous market research products and services."[11] This statement must stand as a masterpiece of indirection. Without stating the name of an individual, one can still convey a number of unique characteristics of that person that, together, distinguish him or her from nearly everyone else. If Uber wanted to exploit data on a rider leaving his or her home, at a given address in a given census tract, calling from a given cell phone number, and heading to a specified destination—those data might well be enough to enable a determined hacker, drawing on data furnished by data brokers, to identify the subject by name.

The information sold by Unroll.me to Uber did not, in fact, include identification by name of those whose accounts had been accessed to provide data for the report—though such identifications can often be reverse-engineered by resourceful programmers. Nevertheless, public reaction could hardly have been more indignant had that identification been included. Lee Tien of the Electronic Frontier Foundation offered a conclusion that all internet users must take to heart: "Many of the services or apps we use for 'free' are monetizing data about us."[12] Other commentators note that even anonymous data about client populations can often be used to identify the people who originated them, should the buyer of the data have a sufficiently strong interest in doing so.

Alternatives

It should be apparent just how prescient the authors of *Records, Computers, and the Rights of Citizens* were in identifying *compartmentalization* of personal data as indispensable for privacy—even if they never used that term.

Note that Reform Four would create a major roadblock to these "finders-keepers" practices. Once any party begins to routinely make and enforce decisions on people on the basis of a standard set of data sources, disclosures from those sources would no longer be permissible without consent from the subject. The routines involved would then constitute a personal-decision system, subject to Reform Four. Though *possession* of personal information would not violate the terms of this reform, the *processing or transfer* of such data for decision-making purposes would certainly do so.

Many businesses and professional practices routinely make earnest-sounding claims that data confided to them will remain "confidential"—an almost infinitely flexible term. Who has not grown tired of vacuous assurances that "your privacy is important to us"—only to go on vaguely to allude to sharing of personal data with "affiliates" and "marketing partners"? But even when an organization actively resists disclosure from its files of personal data confided to it, we must remember that organizations are no less mortal than human beings. From halfway houses to law firms to gynecologists' practices, companies and firms do go out of business, sometimes selling client records or other assets bearing personal information. Entrepreneurs may find it easy to identify interested parties willing to "monetize" those leftover assets.

To some readers, Reform Four may seem too categorical—blocking release of personal information that consumers may not experience as intrusive, or that may even open access to needed products and services. But there is nothing to stop any owner of any personal-decision system from polling those listed in any database and requesting permission to sell or trade their data to specific outside parties. In the absence of any affirmative "opt-in" to such an invitation—that is, advance consent from the subject—no sharing should occur. Any effort to take privacy seriously demands at least that much.

The net result of this far-reaching reform would be something close to the now-standard EU practice in these matters, which permits no "secondary release" of data held in personal-decision systems, except with the subject's consent or under other closely specified conditions. These conditions include (1) a contractual obligation to share such data, as when a consumer agrees to reporting results of a medical examination to a life-insurance company; and (2) legal requirements to report—as in physicians' obligation to report gunshot wounds. Such strictures should hardly block legally mandated retention and sharing of data legitimately held in file. Keepers of personal-decision systems could be required by statute to record and disclose even highly sensitive information—like the existence of a record of offenses against children—to support screening of applicants seeking employment in schools, for example. Nevertheless, the new regime proposed here would put an end to the "free market" in personal data that encourages the aggressive accumulation of such information on spec.

Some exceptions will be necessary. It goes without saying that court orders will sometimes be issued seeking immediate action—as when data held in file are required to initiate "hot pursuit" of a dangerous fugitive. Similarly, legislation should permit disclosures from private sources to protect parties in notably vulnerable situations—for example, to prevent airlines from hiring pilots who are subject to sudden seizures. But the burden of establishing such exceptions must lie on the shoulders of elected officials. One hopes that they will reflect deeply and critically before assenting.

It doesn't take a political scientist to foresee the response to these proposals from First Amendment fundamentalists. For some, restricting the sale or trade of personal data held by corporations or other commercial interests is hardly less acceptable than blocking voters' expression of their political convictions or restricting citizens' rights to assemble for peaceful protest. To quote UCLA law professor Eugene Volokh, among the most articulate exponents of this view, "the right to information privacy... is a right to have the government stop you from speaking about me."[13] In this libertarian vision, protecting the details of people's tax returns from public scrutiny belongs in the same category as suppressing demands for redress from high taxes or prohibiting criticism of elected officials.

But as Volokh would probably acknowledge, restrictions on communications of true personal information in American law are anything but scarce. Bankruptcy law limits how long personal declarations of bankruptcy may be reported by credit agencies—normally to seven or ten years. Personal information on race and gender are excluded in screening for credit, under the Equal Credit Opportunity Act (1974). At the state level, rape shield laws restrict communication of rape victims' identities in media reports. Laws against "revenge pornography" curtail transmission of sexual images recorded in the context of certain intimate relationships. Rights of freedom of expression, like those of privacy, are hardly absolute.

A second major source of organ-rejection-like reactions to this reform will likely invoke a defense of Innovation. Why, spokespeople for America's information industries will ask, did the world's landmark companies of the late twentieth century—Google, Facebook, Amazon, Apple, Microsoft—originate in the United States, rather than in Europe or elsewhere? The historic feats they accomplished require an atmosphere open to new thinking—which requires freedom from privacy constraints. If we want more such corporate golden eggs, the proponents warn us, we dare not make the entrepreneurial ventures of these wondrous geese dependent on the whims of those whose data they use. Everyone, after all, will stand to benefit from the rising tide that will lift all boats in a bountiful sea of personal information.

But the notion that innovation *in general* must always represent a step toward a better world hardly withstands examination. Innovation may create an internet capable of pointing us to products that fulfill our projected "needs"—even in cases where we had been unaware of having such needs. But other innovations generate experiences like that of ninety-two-year-old Richard Guthrie, who was plundered of his life savings (see chapter 1). That feat of technologically abetted larceny was accomplished through a sophisticated scam that began with the purchase of his personal details from a data broker. Still other forms of innovation set the stage for the kinds of invasive data collection discovered by Brian Chen—ingenious means for capturing quite intimate information from what consumers likely consider private communications with their pharmacists, VPN services, health care providers, and the like. It is hardly surprising that these systems respond to the interests of parties paying for them—or that, once up and running, they continue to leave their "fingerprints" on people's lives.

Disclosure from Government Sources

In private-sector personal-decision systems, participation is at least formally voluntary. But inclusion in government systems is more often legally required. Because we may have no choice but to furnish personal information to government agencies—in taxation, environmental regulation, zoning, driving, or a host of other realms—disclosure and sharing of such data must require special justification.

Yet state institutions compile and market vast amounts of personal data—including information whose disclosure is anything but welcome to the subjects. Much of this recording and sharing occurs as an *indirect effect* of the sheer existence of state personal-data systems. The Internal Revenue Service makes carefully qualified promises of secrecy regarding individual income tax returns, for example. But this fact does not stop countless institutions from requiring copies of applicants' tax returns in connection with other transactions—from evaluating job applicants to weighing requests for scholarship aid at colleges and universities. The applicant is, of course, formally free to decline to produce the sought-after documentation. But exercising that option will normally torpedo the application.

Similar observations hold for other government-generated personal documents, such as driver's licenses and passports. The very fact that such crucial forms of personal data are known to be documented *somewhere* grabs the attention of organizations that have their own uses for such data. What foreign destinations has this person traveled to in the past few years? Has this driver had any offenses or accidents officially attributed to him, since his insurance was last renewed? Why is a phone subscriber with no known connections to the Middle East suddenly making and receiving calls to and from that part of the world? The very existence of the documentation in question may make it practically impossible for the subject of the files to resist demands for their disclosure.

Elsewhere, state agencies widely collect personal information—including some very sensitive facts—with the overt intent of making them public. Court proceedings, both civil and criminal, often require disclosure of excruciating material: details of debacles in human relations, bad faith among once-friendly parties, and shared intimacies that will never again be secret, once aired in court. With some exceptions, such court proceedings are subsequently accessible by interested parties. And in perhaps the ultimate exercise of government powers to undermine privacy, those convicted of certain crimes may find themselves featured on state sex-offender registries.

Another powerful institutional motive for disclosure of government-held personal information is money. The ability to compel the collection of vast amounts of data on people's driving habits, criminal histories, court appearances, property ownership, business activities, and the like is unique to government agencies at all levels. Its existence creates enormous demand from companies and other interested parties—where it often fuels decision-making on matters ranging from whom to hire to what terms to offer for insurance or credit. Indeed, among the biggest flows of such state-collected personal information is that serving auto insurance companies. With their intelligence on each driver's history of reported accidents and citations, the states are in a unique bargaining position vis-à-vis the industry, and their sales of such data are a significant source of revenue.

As table 1 shows, state motor vehicle departments can be quite resourceful in their use of personal data that they collect. In Texas, interested parties of the most various kinds can access such data, often for purposes not obviously related to driving. According to a 2013 investigative report by a CBS News affiliate, the tax collector in every Texas county maintains records of data required by the state on what cars are registered to whom. In 2012, nearly twenty-five hundred businesses and other organizations made purchases from this database—including towing companies, banks, hospitals, schools, city governments, and private investigators.[14] Such state retailing of personal data appears to be widespread.

An investigation by Colorado news media in 2018, for example, detailed the case of Eric Meer, a small-business owner working out of his home near Denver. Meer came to suspect that the state was selling his data when, after registering his company with the Colorado secretary of state, he was deluged by direct mail. He found that many of the ads he received were deceptive, asking him to pay fees that he wasn't required to pay. Meer had a hunch the state was selling

TABLE 1 Sales of Personal Information from Seven State Motor Vehicle Agencies

	Most common report	Cost of most common report	Number of most common reports sold
California	N/A	$2.00	N/A
Florida	Name, date of birth, gender, race, height, driver's license/identification card number, mailing address, residential address, and driver history	$8.00–$10.00	8,512,757
Georgia	Name, address, license information, citations, suspensions, and driver history	N/A (cost determined by time and effort)	N/A
Illinois	License information, convictions, suspensions, and accidents	$12.00	3,962,669
Oregon	Name, accidents, diversion agreements, and convictions	$1.50–$8.83	1,085,356

(continued)

TABLE 1 *(continued)*

	Most common report	Cost of most common report	Number of most common reports sold
Texas	Full name, date of birth, driver's license/ID number, issuance date, most recent address, violations, crashes, and enforcement actions	$2.50–$20.00	10,990,317
Washington	Name, address, license information, citations, and suspensions	$13.00	25,208,480

Note: Table 1 was produced by URAP student Pooja Bale in 2020. Entries in the cells were supplied directly from the motor vehicles departments of the states named in the table. In most cases, Freedom of Information requests were necessary to secure release of the information provided here. Most DMV offices contacted by Ms. Bale forwarded the information in documentary form, but follow-up inquiries by phone or email were sometimes necessary. The most frequent buyers of reports were insurance companies, but some states reported selling driver data to many other categories of buyers, as well, including "rental car companies" (Georgia), "vehicle dealers" (California), and "database companies" (Texas). "Withdrawals" are suspensions, revokings, or cancelations of a driver's license and privileges (abbreviated in the table to "suspensions").

his business information to marketing companies. The CBS TV affiliate in Denver confirmed that his hunch was right. Andrew Cole, a spokesman for the secretary of state, insisted that the fees charged to businesses only covered the costs of running the databases. "We are not looking for money," he said. "We charge to cover our costs." According to Cole, there is no way to opt out of these lists and anyone can buy them, even scammers. "There is no screening process," he added.[15]

Here as elsewhere, the organizations creating these record systems have left fingerprints on their creations, as other considerations trump privacy concerns. The resulting commercialization of personal data collected under state authority is anything but an imperative of Technology. Today's sophisticated information technologies, in fact, could just as well serve to implement rigorous compartmentalization of government-held personal information as to disseminate such data. But turning things around in this way would require sponsorship from political forces that are just beginning to assert themselves.

· · ·

True, privacy isn't everything. Every effort to uphold that value involves weighing its claims against those of other, often compelling, public values. State agencies, like private organizations, constantly face demands for actions and expenditures that are costly to the public. Tax collection bureaucracies or administrators of welfare benefits could hardly justify their work, if they accorded tax refunds or family-assistance payments to citizens simply on the basis of the latter's accounts of their situations. Criminal histories, drivers' histories of citations by the police, and property owners' histories of compliance with environmental laws all represent uses of personal data that may not spell good news for the citizens named in them. Yet there are serious reasons for governments to collect such data and, sometimes, to disclose them. Seen in this light, literal and exclusive reliance on the precept from *Records, Computers, and the Rights of Citizens* seems at best incomplete, if not naive.

But such a conclusion hardly supports a diametrically opposite approach, by which government agencies take an entrepreneurial role in seeking markets for data they collect. Better to start with the compartmentalization of personal data collected by government agencies as the default condition, and fashion policy from there by reasoned exception.

Banning the Box and Protecting the Innocent

Some government-held personal-data systems are *designed* for public disclosure. Data from these systems serve government in its role as *disciplinarian*—that is, for purposes of publicly upbraiding, chastising, penalizing, or shaming those considered to have broken the rules. The actions involved range from public disclosure of unsafe conditions in buildings or businesses, like restaurants or health spas, to court uses of criminal records in sentencing and parole recommendations. In such cases, violation of one's privacy through public announcement amounts to part of the penalty. Some jurisdictions even go to the lengths of archiving names and addresses of prostitutes' clients—threatening disclosure to the community at large, in hopes that public naming and shaming will reduce this illegal but irrepressible form of commerce.

No doubt, the most consequential of these systems are criminal records. They furnish the raw material for decision-making on sentencing and parole, as courts and parole boards seek to tailor their decisions to the recorded details of past misdeeds. Perhaps still more important is the role of criminal records in access to employment. Job applications throughout the country commonly include a place for the applicant to indicate whether he or she has ever been convicted of a felony. Answering in the affirmative will, at the very least, complicate most employment applications, if it doesn't bring them to an immediate halt. So pressing are the concerns of employers in this respect that an industry has grown up devoted to "background checking" applicants to ensure that no misdeeds slip under the screening radar.

Thus, the stakes are high, and many voices have risen to challenge the justice of these fateful surveillance processes. Sharpening public concern and contention over these matters is this country's heavy reliance on imprisonment—and the tangled interconnections between that reliance and race. As of 2021, the United States has the largest prison population in the world, with some 2.1 million persons behind bars—a larger total than that of China, the runner-up in this grim competition, which has 1.7 million prisoners in a national population more than three times larger than that of the United States.[16] That amounts to 639 prisoners per 100,000—compared, say, to England, with 131 per 100,000.[17] As of 2021, the United States incarcerates a higher proportion of its residents than any other country. As of 2020, 2 percent of all adult men have spent time behind bars.

The heaviest of these burdens fall on America's Black population. Among Black men aged thirty to thirty-nine, some 5,000 per 100,000 have been incarcerated at some point.[18] As of 2018, more than 1 percent of adult males living in the United States were serving a prison sentence of more than one year—1,055 per 100,000. The imprisonment rate for Black adults was 1,500 per 100,000 at year's end in 2018. And the experience of Black males is most extreme at the bottom of the labor market: among Black male workers who have not completed high school, nearly 70 percent of Black male high school dropouts in their early thirties have served time in prison.[19]

Given these foreboding figures, Black job seekers often face a mind-set among employers that conflates Blackness with high risk.

The net impact of denying employment—or, at least, desirable employment—to every male member of the labor force who has ever been convicted of a felony is chilling. Such practices would, in effect, consign a major chunk of the population to a secondary labor market, outside the normal channels of job recruitment, where the only positions available are those unwanted by the majority. Many observers would characterize the situation in the United States in exactly these terms.

Many spokespeople for Black communities see in these conditions the very forces preventing Black progress in the struggle for racial equality. The prejudice against Black applicants in the search for a better life, they hold, is manifested in "the box" that applicants are asked to check if they have a criminal record. In the words of activist Dorsey Nunn:

> The American criminal justice system has been and continues to be fraught by racism. People of color in general and black people in particular are being stopped, detained, questioned, arrested, convicted and sentenced more often than whites. Black men are six times as likely as white men to be incarcerated during their lifetime. . . .
>
> In 2003, "All of Us or None", an organization I co-founded, started the "Ban the Box" campaign to give formerly incarcerated people a better chance at getting a job. . . . Twenty-one states and 100 cities have banned the box. The sky has not fallen and no new crime waves have hit—it's just people working.[20]

Since Nunn wrote these words, the numbers of states with "Ban the Box" statutes has grown to thirty-four (as of 2021).[21]

Some observers see in the campaign that Nunn helped instigate something almost equivalent to the civil rights movement of the 1950s and '60s. But others have raised misgivings about the repercussions of categorically blocking disclosure of criminal history information. From any point of view, the skeptics hold, the category of ex-felons must include some people who have done some very destructive things—both to property and to people. The owner of a business contemplating hiring from this labor pool must think about two kinds of risks. First, obviously, is potential damage to the business or its employees—as when an employee steals property, sabotages operations, or attacks other employees or customers. Even employers who have, in the past, supported selective employment of the formerly incarcerated may draw back if such an employee misbehaves. The idea of alienating either members of one's staff or one's customer base may leave such employers in a cautious frame of mind.

At the same time, employers must reckon with the tort of *negligent hiring*. This legal doctrine views the employee as the agent of the employer, such that the employer is largely, if not entirely, responsible for the employee's actions on the job. This principle may sound draconian, but there can be no doubt that destructive employees impose *costs* on a number of parties, including customers; the law seeks to apportion these costs.

Thus, this issue poses a conundrum for employers. If their businesses are located in a jurisdiction that categorically "bans the box," they are obliged to avoid any inquiry into applicants' criminal histories. But if they hire without such an inquiry, and the new employee causes damages that might have been predicted on the basis of his or her criminal history, the business becomes a possible target of lawsuits. There can be no doubt that tort actions for negligent hiring can lead to major costs. For example, Reuters reported the case of a car salesman who took a prospective buyer for an extraordinary test drive that ended with the car flipped over and the shopper dead. The employee turned out to have had a long history of "reckless driving, DUIs, and harassment, making him ill-suited to conduct test drives," according to the report.[22] *Ill-suited* hardly seems a strong enough term. This salesman should never have been hired for the job, and the events that disqualified him were a matter of public record. Here the employer could face serious exposure to loss in a negligent-hiring lawsuit.

Perhaps still more perplexing, some research suggests that enacting Ban the Box legislation may itself trigger perverse consequences. In a kind of field experiment, researchers sent job applications to employers in New York and New Jersey on behalf of fictitious candidates, just before Ban the Box statutes were to take effect in the two states.[23] Some of the applications were edited to give the impression that the candidates were Black, while others simulated a white candidate. In addition, some candidates' statements acknowledged a criminal record, whereas others did not. The results were striking: "Before the regulations took effect, candidates with criminal histories were far less likely to be called back, irrespective of race," wrote Harvard economist Sendhil Mullainathan. "After the regulations took effect . . . things changed. Lacking the ability to discern criminal history, employers became much less likely to call back any apparently black applicant. They seemed to treat all black applicants now as if they might have a criminal past."[24]

The framing of this proposed reform's demands may undercut its chances of success. The movement defined its position against inquiring into job candidates' criminal records in categorical terms—opposing the notion that applicants must either check or not check the box acknowledging the existence of such a record. The act of checking conveys no information about the nature of the criminal offense or offenses that the applicant might have on his or her record. Yet many employers, I hope, would draw major distinctions as to the nature of the particular criminal violation in their hiring decision. Some may feel comfortable appointing a person who acknowledges a conviction for larceny, for example, while refusing employment of someone formerly imprisoned for rape. From the standpoint of a movement seeking to build solidarity among its supporters, no doubt it makes sense to present a united front on behalf of all those excluded from employment by recourse to "the box." But the binary choice makes no room for nuance.

Still, public support for the Ban the Box idea continues to spread. As noted above, thirty-four states had some version of this measure on their books by 2021.[25] Given the conflicts and contradictions noted above, however, there are usually qualifications attached. Among the most common is language that permits any employer to reject an applicant because of his or her criminal history *only after* an interview with the applicant—presumably so that employers cannot follow a policy of rejecting all applicants who report criminal histories. Such measures are obviously intended to encourage good behavior among employers—that is, to encourage them to meet such job applicants and not reject them out of hand. But in the court of public opinion, I doubt that any law can ultimately prevail that imposes costs from criminal behavior on vulnerable and innocent figures little suited to withstand those costs. The Ban the Box movement obviously draws from a widespread desire to reverse the systemic losses imposed on Black workers and their families by high rates of incarceration. But Ban the Box in its pure form (eliminating criminal record screening altogether) could impose some terrible costs on other vulnerable parties—for example, women staff members obliged to work at close quarters with those convicted of violent offenses against women. So long as employers can be sued for hiring a former criminal who essentially repeats behavior that led to an earlier conviction, some avenue must be left open for the employer to avoid such suits.

Should states and other government institutions make it their business to document citizens' encounters with the criminal justice system—and to make such knowledge publicly available? Surely some such activities are basic to what nearly all citizens expect of state institutions. States have many functions, but one of these is certainly to act as the voice of public conscience. Just as we have government agencies to protect us against fire and flood, or disease, or foreign aggression, we reasonably expect states to protect us against dangerous people. Such people certainly exist, such that a measure of surveillance is necessary, along with some forceful steps to make recidivism as difficult and unlikely as possible.

But questions of degree demand consideration. Nearly everyone can agree that surveillance in these matters can be too intense, to the extent of stigmatizing offenders indefinitely for infractions that are likely soon to be forgotten and unlikely to be repeated. Policies of limiting the costs imposed for specific offenses aim at avoiding polarization of the larger community. Still, some cases would cause nearly everyone to balk—for example, a criminal convicted of violent offenses against children seeking employment in a school, or a convicted rapist seeking appointment as maintenance supervisor in a women's residence. Or, for that matter, an unreconstructed arsonist in a wholesale paper business. Or a car salesperson with a string of reckless driving convictions, entrusted to take customers on test drives.

In cases like these, I see no choice but for government to defend the rights of vulnerable members of the public against those convicted of criminal acts in settings that would be *substantially replicated* in a new job situation. Tort law seems to have followed this direction, awarding damages in cases of hiring choices that plainly ran such a risk. But how close is *too* close? No doubt, everyone has strong gut feelings on these matters. But we hardly have a formula that could generate a bright line that everyone could agree to recognize.

In a searching discussion of these questions in a law review article, Stacy A. Hickox reports some extraordinary defenses of extraordinary hiring choices by employers:

> Even employers of employees who engage in sexual assault may be able to escape liability in some courts. . . . For example, an employer

was not liable when an employee raped a co-worker, despite the employee's history of using crude and offensive language towards the victim, where the words "did not clearly and unmistakably threaten particular criminal activity that would have put a reasonable employer on notice of an imminent risk of harm to a specific victim. . . . Liability was inappropriate since "not every infirmity of character" is sufficient to forewarn the employer of its employee's violent propensities.[26]

Many spokespeople for vulnerable groups would have trouble with such a formulation. Clearly, legal doctrine on this topic has a long way to go before a consensus emerges that can satisfy all parties concerning how much disclosure, and what form of it, should be required in cases like these.

The Privacy Act of 1974

The closest this country has come to European-style privacy legislation—applying a single set of principles to an open-ended variety of cases—is in the Privacy Act of 1974. These principles derive from the 1973 *Records, Computers, and the Rights of Citizens*—source of the foundational Fair Information Practices (FIPs) discussed in the preface.

The Privacy Act was never intended to apply to the private sector, but only to data held by agencies of the federal government and certain federal contractors. Drafted as part of a national reaction against threats to privacy during the Nixon administration, the bill that became the Privacy Act was one element of a slow-motion collision between deep-going privacy concerns in public opinion and the realities of government use of personal information. This period registered a high-water mark in popular distrust of public institutions. Fueling privacy demands were popular reactions against IRS harassment of individual taxpayers on the Nixon administration's "enemies" list and other privacy-eroding, White House–inspired gambits of the time.

By the spring of 1974, a coalition of political liberals and small-government conservatives was drafting a bill. The bill faced resistance from the beginning, especially from the CIA and law enforcement agencies. But this was a bad time for any public figure to appear soft on efforts to protect privacy. The new administration of Gerald Ford provided its support, and the Privacy Act passed in the final days of 1974.

The Act requires details of all record systems covered by it to be published in the Federal Resister. It establishes a crucial series of rights: a right to *access* one's own records; a right to *amend* certain of these records, if they are found to be inaccurate, irrelevant, untimely, or incomplete; and a right to *sue the government* for violations of the statute—for example, for unauthorized disclosure from agency files—all applications of the FIPs drawn from *Records, Computers, and the Rights of Citizens*. The text of the Act begins with a sweeping declaration, drawn directly from that report: "No agency shall disclose any record which is contained in a system of records by any means of communication to any person, or to another agency, except pursuant to a written request by, or with the prior written consent of, the individual to whom the record pertains."[27]

If statements of general principles made the law, the Privacy Act would amount to something close to total victory for its original sponsors. But from its strong beginning, the Act devolves into a litany of exceptions, qualifications, and reversals. It exempts from its coverage a broad swath of record systems maintained by agencies deemed to be involved in espionage, counterespionage, and law enforcement. The Act mentions the CIA by name, but it also explicitly bars from coverage record systems "maintained by an agency or component thereof which performs as its principal function any activity pertaining to the enforcement of criminal laws, including police efforts to prevent, control, or reduce crime or to apprehend criminals, and the activities of prosecutors, courts, correctional probation, pardon, or parole authorities."[28]

This list of "general exemptions" goes on at length. Its terms are protean. Activities designated as "efforts to prevent, control, or reduce crime," for example, could well be understood to include surveillance measures targeting populations understood to be held *likely* to engage in crime, even in the absence of any actual wrongdoing. It could mean the very sort of unregulated government surveillance of anyone and everyone that generated demand for the Privacy Act in the first place. Readers will recall the discussion at the end of chapter 2 of the highly successful efforts of the FBI, since 2017, to place the contents of its massive Next Generation Identification system off-limits to Privacy Act safeguards. The significance of this step is far-reaching: it means that ordinary, law-abiding citizens required to submit their identifying data to the system (in applying for background checks, for example) are subject henceforth to FBI surveillance—without the subjects of the data being able to inform themselves of such tracking.

A second serious weakness of the Act is its notorious "routine uses" exception. After beginning with the ringing statement quoted above, seemingly proscribing unauthorized disclosures from government-held data systems, the Act goes on to permit release of personal data for "routine uses"—with *routine use* defined as "use . . . for a purpose which is compatible with the purpose for which it was collected."[29] But what makes a use "compatible" with the original collection of data is almost entirely in the eye of the beholder. Is use of Social Security and IRS record systems to enforce court-ordered child support obligations a "compatible" use of those systems? They are certainly used for these purposes today.

The final, and most serious, problem with the Privacy Act is rooted in the logic of FIPs themselves. These ground rules for acceptability in the handling of personal data place *procedure* over *substance*. They posit that, if certain conditions are met—if personal information is accurate, if it is open to inspection and correction by the subject, if the data are used for the same purposes for which they are collected—then the acceptability of the personal-data system is established. Such thinking has every virtue but that of telling us whether the operations in question should occur *in the first place*. It ignores the fact that sheer *access* to personal information is, in many contexts, a highly contestable matter, such that the very existence of certain kinds of databases alters relations among the parties involved. A salient case in point is the recording of prewar Dutch families' religious identifications, which took on life-and-death significance once a Nazi regime occupied the Netherlands.

Or recall the hypothetical case set out in the preface—a government whose legally enacted codes grant law enforcement agencies the right to impose virginity tests on any unmarried woman suspected of having sexual relations. Suppose that such tests were carried out in such a way as to provide those subjected to them full access to the resulting records, and options to challenge the results, if they appeared misleading or incomplete. Would such scrupulous respect for FIPs provide adequate defense against charges that governments simply have no legal basis for such investigations?

The Brandon Mayfield Case

Just how badly things can go wrong when government officials feel themselves freed of accountability in their handling of personal data became apparent in 2006, in revelations emerging from the Brandon Mayfield case:

The U.S. Justice Department said Wednesday it is paying $2 million and apologizing to an Oregon lawyer wrongly accused of being involved with the 2004 train bombings in Madrid, Spain.

Brandon Mayfield was arrested in Portland on a material witness warrant in May 2004, less than two months after the bombings.

According to an FBI affidavit at the time, his fingerprint was identified as being on a blue plastic bag containing detonators found in a van used by the bombers.

The FBI's fingerprint identification was wrong, however, and Mayfield was released several days later.

The bombings of four commuter trains March 11, 2004, killed 191 people and wounded more than 1,800.

Mayfield charged he was a victim of profiling because the Portland-area attorney was a Muslim convert.[30]

The FBI denied targeting Mayfield because of his faith. But it ultimately acknowledged what amounts to flagrant overreach in its efforts to convict Mayfield. The settlement "includes not only a $2 million payment and an apology, but also an agreement by the government to destroy communications intercepts conducted by the FBI against Mayfield's home and office during the investigation."

The written apology reads:

"The United States of America apologizes to Mr. Brandon Mayfield and his family for the suffering caused by the FBI's misidentification of Mr. Mayfield's fingerprint and the resulting investigation of Mr. Mayfield, including his arrest as a material witness in connection with the 2004 Madrid train bombings and the execution of search warrants and other court orders in the Mayfield family home and in Mr. Mayfield's law office."

A Justice Department statement released Wednesday said Mayfield was not targeted because of his Muslim faith and that the FBI had taken steps to improve its fingerprint identification process "to ensure that what happened to Mr. Mayfield does not happen again."

"Mr. Mayfield and his family felt it was in their best interest to get on with their lives," said Mayfield's attorney, Elden Rosenthal.

"No amount of money can compensate Mr. Mayfield for being held as a prisoner and being told he faced the death penalty [for the Madrid bombings]."

Mayfield said his suit was not about money.

"It's about regaining our civil rights, our freedom and most important, our privacy," he said.

He and his attorneys said the settlement will allow him to continue the portion of his lawsuit challenging the constitutionality of the Patriot Act.

Mayfield contends that his home was searched under provisions of the Patriot Act.[31]

The draconian campaign against Brandon Mayfield had gone forward not only under the authority of the FBI and the Justice Department, but also under that of the FISA Court (established by the Foreign Intelligence Surveillance Act of 1978 to provide oversight of intelligence agency activities). For a time, it appeared that everything was falling in line for the FBI's plan for the case. According to a review by the Justice Department's inspector general, Mayfield was arrested on May 6, 2004. On May 17, the federal court hearing the case ordered "an independent expert" to review the FBI's fingerprint identification. "On May 19, the independent expert concurred with the FBI's identification of [the print taken from the Madrid bombing] as being Mayfield's fingerprint."[32]

One wonders where Mayfield would be today, had it not been for a single discordant note—impinging from totally outside the systems of American justice. The Spanish police, always skeptical of the attribution of the fingerprint from the bombing to Mayfield, determined through its own investigations that the print in fact had come from an Algerian national known to their investigators. On May 24, after reviewing the Algerian's prints provided by the Spanish police, the government dropped its case against Mayfield. Every fail-safe device that should have blocked this false (and, in fact, implausible) misuse of personal data—internal review of the case by the Justice Department, the judgment of the FISA Court, and the report of the "independent expert" as to the authenticity of the fingerprint—had failed.

This was not an injustice that the Privacy Act was designed to forestall—no federal systems of personal records were involved, after all. But the Mayfield debacle makes it clear that the climate of opinion that created the impetus for the Privacy Act, in the wake of similar high-handedness by federal investigators, has not succeeded in creating an internal immune system providing effective protection against such grave misuses of personal information.

Dark Regions

Physicists have long held that our familiar material world has a direct counterpart in the form of dark matter. Dark matter replicates its alter ego—with dark protons and other atomic particles, dark worlds, even dark stars. But dark matter defies the attention of researchers. It is difficult to reproduce in laboratories and far more resistant to study than its more familiar counterpart.

Researchers find themselves in much the same position when they address this country's many off-limits personal-decision systems. The vast dark world consisting of systems classified as secret and thus exempt from accountability under the Privacy Act shows only indirect hints and traces of its own existence. Many such forms of tracking and personal-data management take place where one might expect—in the FBI, the CIA, the NSA, and the like. But others occur in less well-known institutions like the federally financed

"Fusion Centers" that have grown up since 9/11 to promote the exchange of personal data by local, state, and regional law enforcement agencies. Personal-decision systems from this "dark side" of American surveillance deserve at least as much attention in any assessment of disclosure and sharing among federal agencies as activities covered by the Privacy Act. Yet they remain largely beyond the reach of researchers' attentions. It is to the discredit of privacy researchers that our writings have focused so heavily on record systems that are openly acknowledged and codified, leaving it to figures like Edward Snowden to draw attention to what appear to be vast reaches of secret recordkeeping that obviously demand attention in the public interest.

If we are determined to take privacy seriously, what changes can we reasonably seek that might redress this situation? Perhaps the single most cost-effective step would be to create mechanisms for broader public scrutiny and awareness of the archipelagoes of personal tracking and decision-making that now comprise the dark matter of American surveillance. We may agree that many of these records deserve to be shielded as part of active investigations of apparently dangerous people and groups—suspected terrorists, for example, or participants in organized crime. But many investigations that gather records of persons suspected of serious criminal or terrorist intent result in findings of no wrongdoing. Are the targets of these inquiries not entitled to know that the invasion of their privacy has come to an end, and that the suspicions once entertained have been determined to be unwarranted?

Many investigations that require court orders for secret surveillance of private data—a suspect's business dealings, for example—require that the target of the investigation be apprised of it. Where prosecutors suspect that such notification could result in destruction of evidence, they may secure an order that permits the search to go forward without notification—followed by notification that the investigation has concluded. This latter step represents an important defense of privacy—for the individual, and for the larger community, to take stock of the reported intent of the investigation and its results.

At stake here is not just the privacy of specific citizens, but something even more vital—the strength of pluralism and of democratic institutions. Without some sense of the frequency of investigations like those into people's bank accounts, personal papers, legal records, and the like, the public cannot hope to develop informed opinion on the importance and legitimacy of such inquiries. This, of course, is one of the elements of the Patriot Act that most disturb advocates of the rule of law: the gag order that it imposes on those served with national security letters requiring them to furnish information and other things related to the person under investigation. Without some means for judging the total extent, and personal impact, of government investigations of these kinds, there is simply no way for the public to assess their justification.

Notorious fiascoes like the Mayfield case demonstrate the weaknesses of current checks on government investigations. Without better scrutiny of such activity, the public has no way to know how common such debacles are. Evidently the parties favoring the gag orders do not want this public debate to happen. They will surely make the case that *any* disclosure of their secret information-gathering activities will somehow weaken the prospects of their "war" on terrorism, criminality, or other nefarious forces. This is the intolerable point for anyone who takes privacy seriously. For us, it is not acceptable to live in a world where anyone or everyone could well be a target of government investigation at any moment without knowing it—regardless of evidence of his or her guilt or innocence. On the contrary, we want everyone to be able to assume that, should he or she come under such investigation at any stage, that fact will ultimately come to light.

None of these aims can be realized, so long as the secrecy of personal-data use by agencies in America's dark regions remains as it is today. To reverse this situation, we need a corps of highly reliable auditors. Thus, Reform Five:

> Create a corps of expert investigators, endowed with security clearances, legal training, and clear authority to report to some form of national ombudsman-like agency or public prosecutors, to monitor uses of personal information in domains presently off-limits. These investigators must be formally qualified to appear before any court, including the FISA Court, and have resources and staffs commensurate with their responsibilities.

This crucial reform would send an indispensable message to the array of agencies now tracking American lives without accountability.

Conclusions and Recommendations

Governments have many legitimate enforcement roles that require cross-checking and corroboration of bases for decisions on citizens—from assessments of tax liability and entitlement to drive a vehicle to checks on people's criminal histories or educational qualifications. Few if any Americans, for example, would support policies that left criminals with histories of financial crimes free to seek employment as personal financial advisors, or drivers with histories of disabling seizures free to take to the roads. Virtually all privacy advocates acknowledge that such authentic needs for discrimination in dealing with people can outweigh privacy demands.

But it will not do to step from this conclusion to an embrace of free disclosure of government-held personal data wherever it may be expedient to or supportive of institutional purposes. Such a policy—already in effect by default in many government operations—would pave the way for exactly what the authors of *Records, Computers, and the Rights of Citizens* appear to have feared: a world where any personal data compiled on behalf of any organization for any purpose become available to any other organization for any other purpose.

The fact that "routine uses"—that is, long-established patterns of disclosure—of personal information from federal files correspond to bureaucratically experienced "needs" hardly settles anything. "Needs" for personal information by parties inside and outside of governments are constantly expanding: the more such data are generated, the more needs for them are discovered. We should return to the essential insight, not as an absolute barrier to the flow of information, but as a *default condition:* compartmentalization of personal data held by government agencies is indispensable for the defense of values that are indispensable to any open society. What that means in practice is that losses to other key values—from efficiency and revenue maximization to the ability to track dangerous figures—must often be accepted as the price of taking privacy seriously.

Notes

1. *Records, Computers, and the Rights of Citizens: Report of the Secretary's Advisory Committee on Automated Personal Data Systems* (Washington, DC: U.S. Department of Health, Education, and Welfare, 1973).

2. Michael Kirby, *Guidelines on the Protection of Privacy and Transborder Flows of Personal Data* (Paris: Organisation for Economic Co-operation and Development, 2013); Michael Kirby, *The Australian Privacy Charter* (Sydney: Australian Privacy Charter Council, 1994); *Model Code for Protection of Personal Information* (Etobicoke, ON: Canadian Standards Association, 1996).

3. Heidi Messer, "Why We Should Stop Fetishizing Privacy," *New York Times*, May 23, 2019.

4. Shoshana Zuboff, *The Age of Surveillance Capitalism* (New York: Public Affairs, 2019).

5. Daniel Solove, *The Digital Person: Technology and Privacy in the Information Age* (New York: NYU Press, 2004), 101–02.

6. Brian X. Chen, "I Downloaded the Information That Facebook Has on Me. Yikes," *New York Times*, Apr. 11, 2018.

7. Julia Angwin and Steve Stecklow, "'Scrapers' Dig Deep for Data on Web," *Wall Street Journal*, Oct. 12, 2010.

8. Tim Libert, "Privacy Implications of Health Information Seeking on the Web," *Communications of the ACM* 25, no. 3 (Mar. 2015): 68–77, 75.

9. Press Release, Federal Trade Commission, "FTC Settlement Puts an End to 'History Sniffing' by Online Advertising Network Charged with Deceptively Gathering Data on Consumers," Dec. 5, 2012, https://www.ftc.gov/news-events/press-releases/2012/12/ftc-settlement-puts-end-history-sniffing-online-advertising.

10. Lauren Silverman, "Turning to VPNs for Online Privacy? You Might Be Putting Your Data at Risk," *NPR*, Aug. 12, 2017.

11. Mark Isaac and Steve Lohr, "Unroll.me Service Faces Backlash over a Widespread Practice: Selling User Data," *New York Times*, Apr. 24, 2017.

12. Lee Tien, Electronic Frontier Foundation, personal communication, Feb. 2021.

13. Eugene Volokh, "Freedom of Speech and Information Privacy: The Troubling Implications of a Right to Stop People from Speaking about You," *Stanford Law Review* 52, no. 5 (May 2000): 1049–1124, 1050–51.

14. "State Sells Personal Information & You Can't Opt Out," *CBS DFW*, February 11, 2018, http://dfw.cbslocal.com/2013/02/11/cbs-11-investigates-your-personal-information-for-sale-you-cant-opt-out.

15. Shawn Chitnis, "People Get Suckered: Small Business Owner Warns Other Coloradans About Confusing Letter," *CBS Denver*, Nov. 26, 2018, https://denver.cbslocal.com/2018/11/26/mailing-list-warning-letter-business-colorado-secretary-of-state.

16. M. Szmierga, "Countries with the Most Prisoners 2021," *Statista*, July 30, 2021.

17. John Gramlich, "America's Incarceration Rate Falls to Lowest Level since 1995," Pew Research Center, Aug. 16, 2021.

18. John Gramlich, "Black Imprisonment Rate in the U.S. Has Fallen by a Third since 2006," Pew Research Center, May 6, 2020.

19. Sendhil Mullainathan, "Ban the Box? An Effort to Stop Discrimination May Actually Increase It," *New York Times*, Aug. 19, 2016.

20. Dorsey Nunn, "Ban the Box Keeps Families and Communities Together," *New York Times*, Apr. 13, 2016.

21. "List of States and Municipalities with Ban the Box Laws," *Accusource* (2022), https://accusource-online.com/list-of-states-and-municipalities-with-ban-the-box-laws.

22. "Can I Be Sued over an Employee's Criminal Record?," *Reuters*, Mar. 9, 2011.

23. Amanda Agan and Sonja Starr, "Ban the Box, Criminal Records, and Statistical Discrimination: A Field Experiment," *University of Michigan Law and Economics Research Paper Series* 16-012 (June 2016).

24. Mullainathan, "Ban the Box?"

25. "List of States and Municipalities with Ban the Box Laws."

26. Stacy A. Hickox, "Employer Liability for Negligent Hiring of Ex-offenders," *Saint Louis University Law Journal* 55, no. 3 (2011): 1001-46, 1021.

27. Privacy Act of 1974, Public Law 93-579, *U.S. Statutes at Large* 88 (1974).

28. Ibid.

29. Privacy Act of 1974 (a)(7), Public Law 93-579, *U.S. Statutes at Large* 88 (1974).

30. Henry Schuster and Terry Frieden, "Lawyer Wrongly Arrested in Bombings: 'We Lived in 1984,'" *CNN*, Nov. 29, 2006, https://www.cnn.com/2006/LAW/11/29/mayfield.suit/index.html.

31. Ibid.

32. Office of the Inspector General, Oversight and Review Division, "A Review of the FBI's Handling of the Brandon Mayfield Case [Unclassified Executive Summary]," (Washington, DC: U.S. Department of Justice, 2006), 3.

4 *Make Personal-Data Use Minimal, Transparent, and Trackable*

Anyone seeking serious reform of personal-decision systems in America today must start by confronting the *asymmetry* of the parties involved. The systems are the work of large, well-organized, and well-resourced government and private-sector organizations. Their opposite numbers in these asymmetrical relationships are us—members of diffuse publics, often skeptical of what other parties are doing with "our" data but lacking ready means for focusing our suspicions and coordinating demands for change.

Without strong grassroots demand to level the playing field on which these ill-matched adversaries confront one another, would-be reformers have often been at a loss. They—we—have often been satisfied with gestures that ultimately fail to equal the magnitude of the forces arrayed against us. These include privacy notices that demand lengthy and sustained concentration, and still leave the reader confused—like the opaque Google user notice quoted in the introduction. Or measures that promise protection of data held in file only on an "opt in" basis—with strong disincentives for choosing this option. Such was the case in planning for the Ontario toll road system included in appendix 2: the privacy option of concealing when and where a driver had used the new highway required an advance deposit of $250. Other allegedly privacy-preserving options include "privacy settings" that baffle many ordinary users and prove very time-consuming to put in place.

In many instances, the inefficacy of such allegedly pro-privacy measures stems from a kind of cognitive overload that they impose on their supposed beneficiaries. Activating safeguards often requires serious user commitments of time and mental energy, in exchange for highly uncertain rewards. Why, one wonders, should I go to the trouble of checking the boxes on a credit card company's privacy notice, and then take the time and trouble to return my work by mail? The instructions I give, even if obeyed, would curtail only a minuscule fraction of the total volume of personal data being exchanged about me. Fighting massive privacy invasion one database at a time, for most consumers, is a prescription for exhaustion and futility—exactly the result that anti-privacy forces seek.

Instead, we need powerful, user-friendly new tools to enable ordinary citizens and consumers to shape the treatment of whole categories of personal data through a few simple directions—or simply by default. And we need, as the default condition, reforms barring information held in personal-decision systems from disclosure—rather than defaulting to the "finders-keepers" principles prevailing in American practice today.

Some critics challenge the need for such measures. The mere existence of today's pervasive data-keeping on ordinary Americans should cause no concern, they argue, without evidence of any resulting *harm*. Did journalist Brian Chen (see chapter 3) suffer *injury*, the skeptics might ask, from the several hundred companies he found tracking him via Facebook? Did these uninvited spectators to his life—apparently acting in the hopes of selling him something, or selling data about him to some other party—really degrade any important interest of his, even if the discovery of their attentions left him feeling uneasy? What reason has anyone to expect the rest of the world to grant him or her privacy, when the absence of calculable damage ought to be enough?

Taking privacy seriously means rejecting this kind of thinking. In dealing with large-scale systems for tracking people and acting on the results of such scrutiny, we need strong positive reasons to justify any such systems—not simply apparent lack of negative ones. We do not normally assign our personal powers of attorney to anyone but trusted parties, for example—even in the absence of demonstrated bad faith or harmful intent from others. Defense of privacy demands similar caution with regard to the potential powers of personal-decision systems. Taking privacy seriously is virtually defined by giving privacy considerations the "benefit of the doubt"—by going to the greatest lengths to protect personal data, given the destructiveness that can ensue when treatment of such information goes wrong.

In fact, tensions underlying this question point to a major divide among ethicists. Much European privacy law takes its inspiration from Immanuel Kant's idea that intrusion across certain intimate boundaries of human personhood is always wrong—even in the absence of physical assault, insult, robbery, imprisonment, or other deliberate cruelty. We rightly recoil, Kantians would insist, at the notion of anyone being forcibly exposed when others remain clothed—just as we do at the idea of anyone's diaries being published without their permission, or anyone's medical record being publicized solely for the amusement of others. Any such actions seem to violate the basic dignity that—Kantians hold—is owed to all human beings. This mind-set underlies what Samuel Warren and Louis Brandeis, quoting Judge Thomas Cooley in their famous *Harvard Law Review* article on privacy, called the "right to be let alone."[1]

Alternatives to this mind-set hold greater influence in the American legal tradition. These competing rationales mostly stem from utilitarian thinking. Utilitarians understand ethical action as resulting from a sort of cost-benefit analysis of alternative courses of action—with the "best" conduct being that which conduces to the greatest total amount of happiness or pleasure. "Very few people want to be left alone," challenges jurist Richard Posner, in a succinct utilitarian manifesto. "They want to manipulate the world around them by selective disclosure of facts about themselves. Why should others be asked to take their self-serving claims at face value and be prevented from obtaining the information necessary to verify or disprove these claims?"[2]

Clearly, that is not the vision that inspires this book. The reasons why we seek privacy are many and varied, I hold, and some of them are indispensable for support of social ties that nearly everyone values. We cherish freedom to refuse disclosure of information about which we feel vulnerable or subject to retribution, for example, or where our opinions are unformed or ambivalent, or where we fear that expressing these states of mind could be hurtful to those around us—or, for that matter, to ourselves. Protections for these tentative or vulnerable sensibilities are not just matters of comfort for the individuals concerned. They represent valuable "shock absorbers" for differences in attitude, worldview, and intent that form part of human interactions everywhere. For these reasons, we should start our reasoning with *no institutional surveillance* as the optimal default condition—leaving justification for watching, tracking, and recording to be established as exceptions warranted by exigent conditions.

Master Portals and Public Declarations

In short, we need a drastic break from the current situation. This state of affairs is characterized by near-universal ignorance among ordinary citizens and consumers as to the origins, movements, and uses of data on themselves—juxtaposed against data managers' extreme sophistication in these matters. In place of this profound asymmetry, we must campaign for a world where every "private" citizen has easy ways of assessing the origins of every item of personal data capable of impacting his or her life—and where he or she can act to help shape those impacts.

We should declare that our current status quo—ordinary Americans' inability to determine where personal information about ourselves is being generated, where it is held, and how it is being used—is tantamount to a national emergency. We must insist on fundamentally new institutions, along with new levers for activating their powers. Accordingly, I propose Reform Six:

> Establish a new national privacy-promoting institution, the Master Portals: two closely coordinated websites offering comprehensive information on personal-decision systems maintained by American organizations, both government and private.

The two portals will provide any member of the public with the equivalent of "one-stop shopping" for information on every personal-decision system declared under the disclosures envisaged in chapter 2. In my vision, the two portals will differ in the information they supply to the public, yet complement one another in their functions. Portal I will furnish a summary of each system's basic operations to any interested member of the public. Based on those declarations, and without disclosing information on any specific person, that summary will include comprehensive accounts, for every registered personal-decision system, of that system's workings. These accounts should include no data about any real-world human being. Instead, they should provide detailed specifications of the forms and amounts of data held on hypothetical people, the typical paths taken by such data among institutional users, and the forms of decision-making that the data ultimately support.

By contrast, Portal II will provide detailed access to all of *one's own* data held in any and all personal-decision systems declared as such to the authorities. Once adequate personal identification is established, any user of the internet should thus be able to determine the state of his or her records, and the potential consequences of those records, as easily as one checks a weather forecast or one's Amazon account.

These powerful and ambitious national tools for privacy awareness and action will require serious institutional support—administration by a regulatory agency like the Federal Communications Commission, for example, or by a yet-to-be-established Personal Data Protection Agency along European lines. No doubt they will require years to be complete in all their intended functions. For the moment, I am less concerned with the identity of the host institution than with the workings of the new portals. They will necessarily draw directly on information included in the public *declarations* required of all personal-decision systems (as described in chapter 2, in relation to Reform One) regarding matters like the system's size, purposes, rates of traffic, and the typical destinations of data generated by it.

In developing these recommendations, like all those in this book, I have sought advice and criticism from colleagues in this country and abroad. One longtime friend and interlocutor, a noted German privacy specialist, minced no words in his opinion of the proposal given above: "All efforts in this direction," he succinctly wrote, "have failed." He identifies the cause of these failures, if I understand him correctly, in the sheer numbers of personal-decision systems—apparently growing every year—and the resulting difficulty, for any administrative agency, of monitoring their activities.

But most of the reforms proposed in this book assume reliance on individual rather than institutional action, at least in the first instance. They aim to create accessible public profiles of what personal-data-gathering organizations keep and what they do with those data. Unlike other commentators on privacy issues, I believe that Americans would be willing to monitor their own data held in file, if provided with succinct directions and effective tools for doing so. Much as people seem to enjoy Googling themselves, I believe that most Americans would show more than passing interest in reviewing the state of their data—and, ultimately, in acting on the results where such action appeared warranted—if only they had confidence that they could turn to an authoritative body for official action, if their investigations indicated they were entitled to some form of redress.

Thus, no registration-and-periodic-inspection regime plays a role in this program. But simply attending to the disclosure requirements given here—especially the public declarations—should nudge ordinary citizens, consumers, *and* data-consuming organizations toward ensuring sharply more privacy-friendly practices.

Reform Seven describes what Portal I must include:

1. The name and business address of the *owners* of the system—or, in the case of government data systems, the agency maintaining the system.
2. The name and contact information of its *keeper*—that is, the party designated as spokesperson for the system, presumably someone available at all business hours.
3. The name of the system (to distinguish it from other personal-decision systems maintained by the same organization).
4. Its *legal bases,* as declared by its keepers.
5. The declared *purposes* of the system.
6. The number of files in the system, the number of persons covered, and the frequency of reporting from the files.
7. Descriptions, using fictional examples, of typical entries in files, including explanations of how to read common codes and abbreviations.
8. An account of typical *uses and destinations* of filed information, for both internal decision-making within the owning organization and the typical outside uses for which data from the system are shared.

This last point is crucial. A key aim of establishing this portal is to enable everyone to track the whereabouts and movements of his or her personal data, as a means of holding data handlers responsible for the movements of information that they are charged to protect.

Thus, any member of the public concerned about any of the personal-decision systems that make up, say, the FBI's National Crime Information Center could access Portal I online and review this information without identifying himself or herself. The information provided on the site might well include hypothetical cases to illustrate the workings of the system under various possible conditions—but never any identifiable personal information on any real person. Simply being able to access this information about any currently active personal-decision system should have a galvanizing effect on public awareness.

Portal II will be dedicated to revealing information from people's own files. The first step to gain entry to this portal will be a process of self-identification, perhaps like the routines that Social Security requires before its offices disclose personal information to callers. Once identification is established, the user should immediately be able to call up a comprehensive listing of *all* personal-decision systems containing files on himself or herself.

Reform Eight describes what Portal II must include:

1. The current contents of one's own file, condensed as necessary, accompanied by a legend interpreting symbols, abbreviations, and the like.
2. A brief account of why information held on file is necessary to fulfill the *purposes* ascribed to the recordkeeping.
3. A brief statement specifying the parties outside the organization that may reasonably be expected to receive the filed information, consistent with the *purposes* ascribed to the recordkeeping.

4. A listing of recent entries to and disclosures from the file, including the dates, the identities of the parties forwarding data and those receiving information, and the nature of the material changing hands.

Note how these requirements support two classic concerns of privacy advocates: first, by restricting filed data to only those data consistent with the system's *legal bases;* and second, by ensuring that the information retained does not exceed what is necessary for the system's legitimate purposes. Everyone involved in managing personal data knows the tendency for files to grow beyond any original intent—perhaps because someone sought to use the extra information in strategic ways, or perhaps simply because of the idea that it's always better to have additional personal information "just in case." Neither rationale justifies retention of information not required for the stated purposes of any system. Keepers of systems would bear legal responsibility for deleting data that do not meet those strict requirements—both in government-sponsored systems and in the private sector.

If We Build It, Will They Come?

This ambitious plan seeks nothing less than a profound transformation in Americans' understanding of, and influence over, the roles of institutionally held personal information in their lives. One can think of the flow of personal data much as one does the flow of petroleum—emanating in great quantity from a number of key institutions and disseminated throughout the population to support all sorts of transactions and relationships. As with petroleum products, we Americans would be well served by developing a mental outline of the creation, flow, and use of personal data in our lives. We would then be in a much better position to debate how much of these things we need—

and what we can afford to do without. The two Master Portals proposed here would satisfy the need for such understanding.

One could argue that Americans, long deprived of such knowledge, have lost all interest in it. But I believe that, once it became clear how enmeshed our lives are in these ongoing flows of data, most Americans would feel the incentive to know more. The very existence of the portals would promote a single set of skills and habits for accessing and managing such information. Today, where understanding of the vicissitudes of data on ourselves is available at all, different steps are necessary to access it from each separate source. Under the reforms proposed here, a single set of commands would provide access to most forms of personal data via the two portals.

Most day-to-day use of the portals, I expect, will serve strictly instrumental, personal purposes. People will want to judge what is known about themselves, who knows it, and what to expect from the organizations liable to access such data. Consider the strictly hypothetical case of Fred, a young researcher in a management consulting firm:

Fred would like to advance his career but is concerned that certain incidents in his past could come to light if he seeks employment outside his present firm. He was convicted fifteen years ago for drunk and disorderly conduct, after an evening of heavy drinking near the college he was attending. Then there is the matter of a personal bankruptcy petition that he filed shortly thereafter, following his father's recent death from cancer. The father's brief but acute illness had drained family finances. Fred has heard from friends and coworkers that any prospective new employer will have the means to bring to light even the most hidden elements of applicants' past histories.

Before applying for another position, Fred accordingly accesses his files in Portal II. His credit files from more than one credit reporting agency show records of the bankruptcy from ten years before. But Fred, on spotting this anomaly, contacts the companies holding these reports to remind them that the latter are outdated. Transmitting outdated information on the bankruptcy, they are aware, could result in a stiff fine. In the case of the arrest record, Fred notes on Portal I that data brokers had once compiled records of this event, but that these organizations are operating without legal bases for that sort of reporting. Unless someone manages to inquire directly to the police for that information, he determines, no intermediary agency will have the power to resell it to a prospective employer. Thus reassured, Fred goes about his quest for a new job without unwarranted concern about any stigmatizing reports from the past.

Think of the pressing desire of Americans to avail themselves of the national "Do Not Call" registry as a shield against "junk" telephone appeals—despite its highly mixed effectiveness in blocking such calls. Or consider the option to "freeze" one's own credit file, so that no reports can be issued from it without one's explicit authorization. Both of these ultimately popular measures were vigorously resisted by their respective industries—yet their adoption rates have been very high. If using the Master Portals were as easy as Googling oneself, recourse to them would be nearly as intuitive.

Remember, the portals will work in tandem with the other reforms proposed here. Where consumers' inquiries reveal retention of personal information without a legal basis—which will often mean without consent from the subject—the latter will be able to communicate directly with the keepers of the systems to demand that such retention cease. Here, I draw inspiration from the success of California's Proposition 24 in 2020, the sweeping privacy measure ultimately supported by some 56 percent of the state's voters. Proposition 24 expanded and amended the California Consumer Privacy Act of 2018, giving California consumers the right to sharply restrict the sale or disclosure of personal data held by large businesses.[3]

Creating these comprehensive new systems of surveillance-oversurveillance will obviously be a time-consuming enterprise. The two portals will require technological innovation and field-testing experimentation to work well. They will require a phase-in period during which administrative practices evolve in tandem with technological refinements. Accordingly, it's likely they will have to be phased in over years, with different elements of the systems added in light of evolving experience.

Will consumers come? This is, of course, a fitting challenge for any proposed innovation in public participation. But I believe that the demand for access to the information and guidance the portals afford will be felt immediately. These sources of pertinent data on matters directly affecting one's own life will be vastly more useful and user-friendly than the balkanized, inconvenient means of informing oneself available today. Above all, the existence of the portals will promote a single set of skills and habits for access that will work for any system. Users will not face a new set of user-unfriendly ground rules each time they set out to track the status and effects of their own records.

One would hardly conclude, from the fictional scenario above, that the reformed system has enabled Fred to escape responsibility from all negative elements of his past history. But Fred has managed to press his own case so that the new entry under his name seems to bear his "fingerprints" at least as much as those of the organization maintaining his file. At the very least, the portal has enabled Fred to put his best foot forward.

But in similar circumstances, not all awkward or negative elements of one's biography would be susceptible to such editing. Consider the hypothetical case of Fiona,[4] who faces concerns similar to Fred's:

Fiona, a single Black mother in her early thirties, is proud of the strides she has made in recent years. Since her mid-twenties, she has worked as a corrections officer, receiving several promotions and becoming a member of her union's local council. She has been nominated to join the negotiating committee for the next round of contract talks, a potential stepping-stone to higher things in the union. These negotiations are bound to receive close coverage in the local media.

But Fiona is uneasy about taking a role in the high-visibility talks. Her union colleagues have warned her that that their employer has engaged professional negotiators for the new contract who are known for attacking the reputations of prominent figures among their union adversaries, hoping to undermine their standing and that of the union itself. Now Fiona worries that the other side may uncover pornographic videos in which she played a role. Via Portal I, she determines that a number of organizations that distribute X-rated videos maintain lists of performers that are searchable. Such videos have already resulted in comments on the local TV station, as well as on social media.

Anxious about what might come to light, Fiona searches Portal II for entries bearing her name. Sure enough, though her total archive of personal records is spare, she does find "her" video marketed under a stage name easily associated with her given name. Hoping to block further circulation of the video, she finds that she signed a contract assigning rights to the footage to the company that produced it. Accordingly, she is unable to block its dissemination. Determined to avoid anything that might bring further publicity to a chapter of her life that she strongly wants the world to forget, and that could weaken her union's position in the coming negotiations, she withdraws from further union roles.

Fiona's setback illustrates some limits of the reforms proposed here. True, the information accessible through the two portals would provide subjects a much fuller grip on which of their personal data are apt to surface, and to matter, in various contexts. And the reforms themselves should limit the amount of recorded information available and the uses to which such data are put. But nothing in these proposals is intended to immunize anyone from any and all negative fallout from past events. We have to assume that voters, legislatures, and courts will continue to deal with cases—like one's appearance in an X-rated video, in this case—that may be considered newsworthy or valuable as precedents for future cases. Still, while the portals would not necessarily enable people in Fiona's situation to escape the repercussions of all the controversial elements of their pasts, they could provide forewarnings sufficient to help shape personal strategies for coping with the possibilities of disclosure.

Responsible Agencies: Enforcement and Precedent

Most of the reform ideas advanced in this book thus far have pointed to steps that ordinary citizens and consumers can take for themselves. Proposals for the Master Portals, for example, aim at mobilizing people's awareness of and involvement in the fate of their own data. Sheer curiosity about what personal-decision systems exist and how they shape one's own life, I suspect, will move many Americans to consult the portals. Where such inquiries uncover data held in systems with no legal bases, or data deemed outdated or excessive for its declared purposes, citizens and consumers will have easy ways of challenging the keepers of the offending system, either to eliminate the data or to secure the subjects' permission to retain it. Perhaps even more basic is the right to "just say no" to inclusion of one's data in any system where not legally required (as detailed in chapter 5). In addition, sponsors' knowledge of significant penalties for breaches of the reformed privacy standards proposed here should make them responsive to individual initiatives of this kind.

But self-activating steps can never suffice as the only mechanisms for making good on these reforms. However minimalist in their concept, virtually all these bodies—let me call them *responsible agencies*—must include in their duties procedures for resolving disputes over treatment of personal data. Yet there is reason to believe that some practices and mechanisms for such dispute-resolution promise stronger privacy protection than others.

Some responsible agencies mediate citizens' disputes over personal data solely in response to requests from the public—most often, where a citizen or consumer objects to treatment of his or her data. In my vision for the reforms proposed here, the portals described above will elicit queries and complaints by ordinary citizens who may otherwise have been unaware of the form and extent of recordkeeping on themselves. Indeed, information conveyed by the portals will often give rise to demands for change—in the content of records, in the very existence of those records, or in actions taken in light of the records. In the most encouraging of scenarios, what people learn via the portals could lead them directly to seek changes from the keepers of record systems, who could hasten to amend filed information without official intervention. But if such steps do not yield agreement, one or both parties must appeal to the responsible agency.

Here we need to ask: What about systems of complaint-driven conflict resolution in situations where grounds for complaint are hard to detect by ordinary consumers? What if, for example, the owners of a website on sexually transmitted diseases (like the hypothetical one described later in this chapter) are found to have engaged in illegal sales of identifiable personal information gleaned from the website to third parties—say, for advertising and marketing purposes? Under circumstances like these, responsible agencies endowed with powers to launch investigations in their own right—without requiring consumer complaints—will be in a stronger position than complaint-based bodies.

Then there are questions concerning what counts as success when the privacy authorities take on tasks of dispute resolution. Consider another fictional case:

Judith, an accountant in her forties, appeals to the responsible authority to reverse a decision by her employer. The large accounting firm where she works has, it seems, declined to consider her for an important promotion, because of data from her employee medical file. These records indicate, the organization claims, that Judith has genetic predispositions to cognitive disorders likely to reduce her abilities to concentrate later in her working years.

Judith insists that the purpose of creating her medical information files at the company had nothing to do with making determinations of this kind. Use of those files to support the company's judgments about her future job skills should, accordingly, be denied. But management insists that its judgments were strictly performance-oriented and that the files developed on Judith were all regarded as legitimate when collected.

In response to challenges like this, responsible agencies in various countries have taken quite different approaches to dispute resolution. One approach might take the form of a mediation exercise:

The agency calls Judith and company representatives together in a closed-door session. After presentation and rejoinder on the facts of the case, the mediator proposes a compromise: Judith will be allowed to apply for the new position—but she must agree to tests of her cognitive function at regular intervals. Should future tests reveal any loss of mental acuity beyond what normally accompanies aging, she will agree to take early retirement, which the company agrees to make available.

The company, little desiring to set a precedent in these matters, and Judith, not wishing to disseminate any grounds for doubts about her thinking abilities, agree to sign a statement proposed by the mediator blocking any disclosure of the deliberations and the decision issuing from them. The statement they sign makes the agreement binding but enjoins the signatories from disclosing anything about it.

One can see how an agreement like this might be satisfying to the two parties—though for different reasons on the two sides. Judith may prefer not to let anyone she knows learn of questions raised about her cognitive health; the company may prefer not to advertise its willingness to grant early retirement as a tool for dispute resolution. But this resolution of the conflict has very different implications for the future of privacy than a second possible approach:

On receipt of Judith's complaint, the responsible agency convenes a hearing on the dispute between her and company representatives. The two parties present their cases, followed by an opportunity for the two, and the agency representatives, to question one another. In addition, several labor unions besides Judith's present arguments, as do several management consultants acting on behalf of the company.

Following the hearing, representatives of the data protection agency withdraw to consider their position. They then announce that they have decided in Judith's favor. Judgments about the supposed future health of an employee, they announce, should never be based on predictive models stretching so far ahead as the one used by this employer, regardless of whether the data on the subject's medical status derive from internal or other sources. A statement of the basic facts of the case and the outcome is prepared and recorded for the use of similar panels in deciding similar disputes in the future.

The contrasts between these two scenarios are profound and full of implications. They invoke contrasting visions of "success" for any responsible agency. The first approach obviously aims at the comfort and satisfaction of the parties, placing main emphasis on finding a compromise that protects some vital interests of each. To accomplish this, the responsible agency encourages the two parties to make the agreement effectively secret. But secrecy is the opposite of what is sought in the second model, in which the aim of the board is to do justice to the claims and counterclaims of the parties, and in which it is crucial to identify a precedent that will relate to as broad an array of future cases as possible. Hence the push to get the details of the study before the public as soon as possible—no doubt relying on pseudonyms for the names of the principals—in hopes that any precedent will serve to support further decision-making.

The power implications of these contrasting approaches are also far-reaching. The first, a mediation-based or counseling scenario, treats disputes as strictly personal matters—as though they took their character from idiosyncratic qualities or transient feelings of the parties. Under these circumstances, the more resourceful party—normally the organization doing the recordkeeping—always has the option of making an attractive offer to the complainant, but with the condition of suppression of the details of the case. Such a step effectively ensures that no one on the outside draws any broader lessons from the dispute; it remains a private matter, with either side free to deny any attributions that might be made against it. The victims of deliberate abuse of filed data, for example, may be required by the terms of the settlement never to repeat their claims.

Americans have grown sensitized to the significance of resolutions like this in the wake of the notorious "Me Too" controversies over sexual harassment. There the (usually male) perpetrators have often offered lavish cash settlements to female employees and contractors who have complained of unwanted sexual advances—always on condition that nothing be disclosed by either party after the signing. Until lately, few of the victims in these proceedings seem to have been willing or able to resist such offers in the interest of retaining the right to go public. The fact that some victims have now made a public example of holding out in just this way may have marked a tipping point, at which the effectiveness and possibility of resistance seems to be becoming a new norm. One can only hope that the architects of dispute resolution in matters of privacy and organizations draw the logical lesson from the parallels in matters of sexual harassment.

The Purposes of Institutions

I believe that implementation of these reforms would foster a kind of "creative destruction" in public life. America's total population of personal-decision systems could well register a major collapse—as all sorts of data brokerages and other personal-decision systems that cannot meet the standards envisaged here either close their doors or fundamentally transform the terms of their dealings with their subjects. As in the fabled meteor that apparently brought an end to the era of the dinosaurs, a shakeup of these proportions might well give way to new forms of institutional life, in a more privacy-friendly environment. Remaining systems might then be subjected to much intensified public scrutiny—if only because there would be more critical attention to go around. Not only the accuracy of the data held in file and the correctness of the decision-making processes of the remaining systems, but even the fit between the data they hold and the purposes they now set for themselves might come under new scrutiny.

But here the conceptual going would get rough. *Purposes,* either of individuals or of complex organizations, are not exactly the open-and-shut, factual matters that we may imagine. The process of discerning, or defining, such purposes is in fact a highly *political* business. But why is that, one might ask. If one wants to know the purpose of any social organization, why not simply *ask* participants in it what it seeks to accomplish? Because the purposes commonly ascribed to institutions—companies, political parties, religious bodies, and so on—are not necessarily identical to those of their participants. Or, at the very least, they're not the same for *all* participants, or even all mere bystanders. From the perspective of an auto worker in a vast assembly plant, the institution in which she works may exist to generate a livelihood for people like herself. From the management's point of view, its purposes may have more to do with giving scope to their own prowess in coordinating a vast and complex organization. For stockholders, the point of all these activities may be to maximize return on invested capital. These observations are not peculiar to any one kind of organization. Every institution, from the corner grocery to the Church of Rome, is bound to include people sleeping in the same bed but dreaming different dreams.

Think of a hypothetical example often mentioned by privacy advocates, and touched on above: a website offering information on sexually transmitted diseases. The creator of this amenity, let's imagine, is a pharmaceutical company that produces drugs to treat such conditions. The site provides reliable information on the symptoms, prognoses, and treatments of various STDs, along with free subscriptions and information on new product developments. The site also requires visitors to register, and in so doing places cookies on their computers so as to track their other interests, information sources, and buying habits. These practices enable the company to pitch its products to visitors to the site, well after their visits, and even in cases where the visitors never wanted note of their visits to go on record.

The indignation likely to be provoked by these practices hardly needs explaining. Disgruntled visitors, if informed, would decry collection of their personal information as a violation of (what they consider) well-established norms, including the confidentiality of medical communications. "People's willingness to seek intimate medical advice," they might complain, "presumes a certain respect for their privacy from caregivers." In this view, capturing such information and using it to hype products to website visitors, without their advance notice and assent, represents a breach of a crucial norm.

But one can equally well imagine the responses of the companies. Creation of the website is a business activity, they would point out. It can be justified only by generating sales sufficient to pay its costs. Without the opportunity to contact potential customers, there would be no website, and hence less information available for everybody. Moreover, the defenders will remind us, neither the website nor any other source—in this hypothetical example—ever provided any assurance about the uses to be made of visitors' information.

Dilemmas like this remind us of a pervasive reality in personal-decision systems that law and policy often seem programmed to ignore. Defining the *purposes* of any enduring social unit is a political step. Here I mean *political* not in the everyday sense of running for public office or taking stands on bills before Congress—but rather in the classic sense of involving collective efforts to define the ends of governance itself. What criteria justify creating a website like this in the first place? Does profitability for the sponsoring company suffice, in itself? Or do such criteria of acceptability need to include some form of contribution to the well-being of the larger community?

By my lights, the best institutions are those that can claim support from the broadest array of such interests. Thus, Reform Nine:

> In the public declarations required of every personal-decision system, the declared purposes of the system must refer to purposes shared across a variety of social roles and groups associated with the system. Those purposes must never merely be the purposes of the parties creating the systems.

In short, to warrant its *legal basis*, any personal-decision system must be acknowledged to serve a variety of widely supported interests and values.

But judging what values or interests can legitimately be considered part of any institution's purposes is rarely an open-and-shut case. Think of the Clinton Foundation, whose announced objectives include improving global health and wellness, increasing opportunity for girls and women, reducing childhood obesity, creating economic opportunity and growth, and helping communities address the effects of climate change.[5] Most observers would have little trouble identifying the foundation simultaneously as a vehicle for the Clinton family's political goals. Of course, both characterizations may be accurate in one context or another. But judging the relative preponderance of such considerations is a matter of subtle interpretation, to say the least.

Similar issues arise in the many cases where privacy codes seek to limit the range of personal information that organizations may legitimately rely on in their decision-making on people. The classic formulation of Fair Information Practices from 1973 (as discussed in the preface) states, in two of its five principles: "There must be a way for an individual to prevent information about him obtained for one purpose from being used or made available for other purposes without his consent"; and "Any organization creating, maintaining, using, or disseminating records of identifiable personal data must assure the reliability of the data for their intended use and must take reasonable precautions to prevent misuse of the data."[6] The operative term here is *intended use*. Whose intent, and what uses, does this language cover? The far-reaching repercussions of this studied ambiguity in American privacy doctrine have not received the attention they deserve. A key case in point is the weakness of the Privacy Act of 1974, in its blanket exclusion from privacy guarantees of what that law brackets as "routine uses" of personal data. The interpretations given to this magnificently ambiguous terminology have permitted government officials to characterize virtually any disclosure of personal data that is *expedient* as *routine*.

Changing understandings of such ordinary terms as *purpose* or *cause* underlie all sorts of explosive differences in practice. Think of the complexities that have arisen in ongoing legal struggles over responsibility for sexual assault and rape in American universities—and their many quasi-judicial counterparts in businesses and other organizations. At one time, information on a female rape victim's sexual history was held relevant in defining whether a crime had even occurred. Today, such questions are intensely contested, on grounds that no history of the victim's sexual behavior can exculpate those who commit rape or other sexual assault. Women are entitled, defenders of the new rules insist, to freedom from sexual abuse, *regardless* of factors like their reputations, their supposed past actions, or what they were wearing. The question is not an empirical one—for example, whether a history of a given form of conduct makes sexual assault more likely—but rather a strictly moral judgment of what constitutes unacceptable conduct.

For a (currently) far less volatile example, consider personal bankruptcy law in the United States, and the personal-decision systems associated with it. If guided solely by strategic calculation, banks, credit card companies, retailers, and other creditors would probably decline credit applications from consumers with any history of bankruptcy—on the theory that those prepared to take this step at one point in their lives are more likely to do the same thing later on. But the law of many states sets limits (generally seven or ten years) beyond which filings of personal bankruptcy may not legally be reported by credit reporting agencies. Here, as elsewhere, what constitutes a "complete record" as a basis for decision-making on the applicant is a matter of social determination.

All this counsels a critical attitude toward data-based decision-making in systems like those studied in this book. What constitutes "completeness" or "all relevant information" on matters like the "intended use" of personal information held in file? Answers to questions like these are not evidenced in nature—like the speed of light or a person's weight. Interpretation of the Privacy Act of 1974 have often been challenged by questions, say, of whether disclosure of personal information from the files of a Social Security enrollee should be considered "compatible" with the purpose of that recordkeeping system. Surely the interpretation of "compatibility" in this context amounts to another deeply value-charged political judgment that cannot be resolved simply through appeal to empirical evidence.

Voting: Its Purpose

To hear them tell it, many public officials in this country have recently encountered a dearth of personal information available for the intended purposes of their work. Those charged with administering elections report themselves unable to register some would-be voters because the officials cannot be certain of the existence and place of residency of people whose names and residence in their districts would, to most of us, seem well attested. Authorities have often felt themselves obliged to cancel voters' registrations because of lingering doubts about crucial personal information retained on file. As one Republican stalwart stated the matter, "an up-to-date voter roll is, to me, [more likely] to give you a purer election, than one that's not."[7]

The systematic efforts of white Americans to suppress the political voices of African Americans go back to the days of Reconstruction, with strategies ranging from physical mayhem and intimidation to present-day database-management legerdemain. In the early twentieth century, Black citizens faced poll taxes that, if unpaid, disqualified them from voting. In the face of widespread public disapproval throughout the United States, the authorities in segregationist districts later turned to "literacy tests" as a requirement for voter registration. These tests required prospective Black voters to respond to questions like "How many grains of sand are on a beach?" and "How many bubbles are in a bar of soap?" White registrants never had to respond to questions like this in order to join the lists of registered voters. "While these laws did not explicitly mention race," comments writer Olivia B. Waxman, "county registrars often applied them unequally, circumventing the law and disenfranchising Black voters."[8] The civil rights movement largely seems to have discredited those tactics. It led to the Voting Rights Act of 1965. This legislation brought federal supervision of voting to a number of states that had engaged in such practices, opening a period that saw major gains in voting by Black residents. But since then, the Supreme Court has curtailed that federal supervision, and the original intent by white authorities is back. Now the tactics are more sophisticated, though the strategy remains the same. It involves computerized purges of Black voters from the rolls—without any need for face-to-face confrontation between public officials and would-be voters.

Among the most concerted (and effective) of these efforts have been the state of Georgia's. In July 2017, Secretary of State Brian Kemp (Georgia's governor at the time of this writing) had some 260,000 voters purged from the state's voter rolls, imposing what later came to be called his "use it or lose it philosophy of voting." These citizens had, in Kemp's view, allowed their status as voters to lapse by failing to vote often enough. Not that this gesture was directed against any particular constituency, of course. Kemp maintained that he was simply trying to protect the state against voter fraud.

Other white-supremacy-minded states have developed their own methods for "keeping the voter rolls up to date"—to use another upbeat, see-no-evil characterization that some purge proponents favor. Nobody really believes that the "integrity" of voter rolls is what's motivating these Republicans—perhaps any more than anyone doubts that Democratic administrations are deeply interested in maximizing the votes of their supporters, when they encourage balloting by mail or extend the dates by which absentee ballots may legally be cast. Higher representation of eligible voters means broadened democracy with a small *d,* though, whereas concerted efforts to minimize participation in the political process by Black citizens serves another set of political values altogether.

But step back for a moment. Here we are, by many accounts, in the midst of an "information society," in which personal data, like countless other forms of information, are available to a degree and in a variety quite unprecedented. The political regimes throughout the country that now devote themselves so relentlessly to cultivating the "purity" or "integrity" of the voting process are not fooling anyone. Nor do they apparently have any need or desire to deceive: their most thick-headed supporters will decode those catchwords in the blink of an eye and embrace the message instantly. Pressed a bit harder, defenders of the purges of voter rolls described above may cite anxieties about voters who have moved, been imprisoned, or died. But evidence that such inevitable variation has made a difference in the *outcomes* of elections remains very scarce.

The only reasonable explanation of the motivation for this trained incapacity of election officials to exploit available sources of information regarding eligibility of would-be voters is political, not technical. And the politics motivating the Republican position lie squarely in the most loathsome traditions of white supremacy. Reliance on seemingly impersonal methods of computerization and algorithms to implement these retrograde purposes should deceive no one. If the purpose were truly to achieve a high level of certainty on the legal eligibility of those seeking to vote, two tax law specialists have noted in the *New York Times*, IRS databases could be much more efficient than local voter registries.[9]

One could also imagine an independent national institute for documentation of the American population, to be tasked with drawing from the greatest multiplicity of sources—both private and public—to answer authoritatively and promptly any question about the eligibility of any particular person to vote. Such a central clearinghouse would provide feedback to local and state registrars about whether a person of a given name lived at a given address over a given period. Its sources could include government bureaucracies—from the IRS to the Selective Service System to the State Department's U.S. Passport service—as well as private credit reporting and marketing databases.

The question is not one of integrity or prevention of fraud—but one of political will.

Notes

1. Samuel D. Warren and Louis D. Brandeis, "The Right to Privacy," *Harvard Law Review* 4, no. 5 (Dec. 15, 1890): 193–220, 195.

2. Richard A. Posner, "The Right of Privacy," *Georgia Law Review* 12, no. 3 (March 1978): 393–422, 400.

3. For details of the outcome of Proposition 24 at the polls, see "California Proposition 24, Consumer Personal Information Law and Agency Initiative (2020)," *Ballotpedia*, Nov. 3, 2020, https://ballotpedia.org/California_Proposition_24,_Consumer_Personal_Information_Law_and_Agency_Initiative_(2020). No statewide statistics exist to show the extent to which California residents have availed themselves of the new rights afforded by these measures. But among major corporations in 2022, Uber reports receiving and complying with some thirty-nine thousand customers' requests to opt out of the sharing of their data, while the Disney corporation reports nearly sixty-three thousand such requests in the same period.

4. For a discussion of the effects of past participation in the pornography industry on the later professional lives of women, see Safiya Umoja Noble, *Algorithms of Oppression: How Search Engines Reinforce Racism* (New York: NYU Press, 2018), 119–21.

5. "Leadership," Clinton Foundation, accessed Dec. 14, 2022, https://www.clintonfoundation.org/about-the-clinton-foundation/leadership/.

6. *Records, Computers, and the Rights of Citizens: Report of the Secretary's Advisory Committee on Automated Personal Data Systems* (Washington, DC: U.S. Department of Health, Education, and Welfare, 1973), 41.

7. Angela Caputo, Geoff Hing, and Johnny Kauffman, "After the Purge: How a Massive Voter Purge in Georgia Affected the 2018 Election," *American Public Media*, Oct. 29, 2019, https://www.apmreports.org/story/2019/10/29/georgia-voting-registration-records-removed.

8. Olivia B. Waxman, "'I Vote Because': Americans Are Sharing Stories of How Their Ancestors Overcame Discrimination in Order to Vote," *Time*, Nov. 3, 2020.

9. Jeremy Bearer-Friend and Vanessa Williamson, "The I.R.S. Can Register Voters as Well as the D.M.V., and Maybe Better," *New York Times*, Sept. 29, 2021.

5 Institute a Right to Resign from Personal-Decision Systems

How much personal information can "society"—the state, the family, the corporation, or any other unit—legitimately demand of anyone? When does a supposedly "private" citizen have a right to refuse pressures to share the details of his or her life—or, indeed, to "resign" from systems that demand personal data? These questions underlie every part of this book—both as an agenda for empirical inquiry and as an invitation to reflect on the requirements of social justice.

Such sweeping questions have no unique, once-and-for-all answers. Context is everything in struggles over personal information, and the contexts of personal data vary infinitely. But in relations between ordinary Americans and well-organized, resourceful institutions, I have long felt that establishing a basic right to withdraw from any personal-decision system not required by law would represent a positive step.

For reasons explained below, I no longer believe that such an unqualified "right to resign" from any and all such systems would make sense. But I do hold that a carefully specified right to withdraw—as a default condition, subject to closely justified exceptions—would mark a distinct gain for privacy. Such a reform could empower those seeking to free themselves from high-handed demands by information-hungry organizations. It would also provide a much-needed nudge to the organizations themselves—which, acknowledging their personal data-keeping practices, might then feel moved to make them more privacy-friendly.

Resigning from Troublesome Listings

At many key moments, being on the right list—say, the short list of finalists for a choice job, or an invitation list for an exclusive event—can be all-important. In these cases, we are normally happy that the other party knows where to reach us and when. But being on the wrong kind of list can be fatal. Recall the story of Onree Norris, a seventy-nine-year-old Black man in rural Georgia (as recounted in the preface). Law enforcement had included his house—ineptly, it turned out—in a list that cast him as a drug dealer. But there was no drug activity at Norris's address. Some drug dealing was suspected at a nearby house of a different description and a different address. The first indication that Norris received of his inclusion on the crucial list was the explosion of a flashbang grenade thrown into his residence by a "dynamic entry" team of law enforcers, who then smashed their way into his house with a battering ram. Norris, who had a heart condition, was handcuffed but was later released. He survived the ordeal. The police who invaded his home faced a lawsuit on his behalf, but the judge hearing the case excused them on the grounds of "qualified immunity" of public servants in the execution of their orders.

Not so lucky was Breonna Taylor of Louisville, Kentucky, also African American and also the target of a notorious "no-knock" warrant. The twenty-six-year-old was known in the community for her work as an emergency medical technician. In this case, the police reportedly suspected that illegal drugs, or perhaps large amounts of cash, were hidden in her apartment. She and her boyfriend, Kenneth Walker, were at home there on March 13, 2020, when Louisville police officers began smashing in her front door with a battering ram. Walker fired one shot, to be met with a fusillade of thirty-two shots from the authorities. At least eight of the police bullets struck Taylor, killing her.

No drugs or other contraband were discovered in the apartment. The police officers who had obtained the warrant for the no-knock raid later pled guilty to falsifying the information they had presented to a court to obtain the warrant.[1] But no charges were filed against the officers for firing the shots leading to Taylor's death.

Needless to say, neither Onree Norris nor Breonna Taylor had the opportunity to scrutinize the lists that targeted them before the violent home invasions that those lists inspired. One has to wonder how much scrutiny the documents authorizing these extreme measures received from anyone in the law enforcement offices that prepared them. If the residents of the neighborhoods targeted had been affluent white people, wouldn't someone have noticed the anomaly involved in launching such an operation against, say, a seventy-nine-year-old retiree?

Grotesque events like these compel reflection on two questions. The first is how such life-threatening misinformation could have found its way into the systems that triggered these actions by incompetent public officials. The second—still more crucial, in my view—is what public value looms great enough to warrant invasions of this kind under any circumstances. Illegal drugs certainly cause suffering and even death. But so do bad diet, hazardous sports, or driving while emotionally distracted or dangerously fatigued. None of these vices, however deplorable, undermines basic premises of an open society, limited government, and rule of law. But the resort to no-knock tactics by the authorities threatens those things profoundly. If the politics of personal information cry out for reform, they do so nowhere more than here.

Public and Private Demand for Personal Data

Most of us accept that some needed government activities require collection and use of personal information. These range from the documentation of vital events—births, deaths, and marriages—to records of eligibility for medical and retirement insurance, to criminal records and the tracking of those reasonably suspected of supporting terrorism. We urgently need public soul-searching and debate on how privacy-friendly these surveillance measures should be, and how severe the response should be when they go wrong. But in any assessment, the potentially fatal no-knock raids represent a step too far.

In the private sector, provision of one's own personal information is normally at least formally voluntary, but often the stakes are so high as to leave the subject little meaningful choice. Applicants for employment are widely asked for access to their credit reports and social media passwords—strictly at their discretion, of course. But everyone understands that refusal to comply is apt to end the discussion and the candidacy. Moreover, private parties often thrive by the resale of personal information purchased from government sources, while at the same time government agencies are major purchasers of personal data from private-sector data brokers.[2] Not all these public-private exchanges give rise to inspiring stories, as an article by Olivia Solon in the U.S. edition of Britain's *Guardian* newspaper makes clear:

> Kim (not her real name), a Canadian in her 30s who runs her own design company, is scared that prospective clients might see her mugshot, which was taken after a violent—and what she regards as wrongful—arrest as she was leaving a gay bar in Florida in 2014....
>
> About a month after the incident, Kim's lawyer called her to say her photo had been posted online. She turned to Google in a panic.

Although her charges were dismissed weeks before, her battered face was staring back at her from several mugshot websites.

"It was mortifying. I'd just started my own company, and my reputation is everything," she said. . . .

She started to pay fees of between $200 and $450 to remove photos. Each time her photo was taken down from one site it would pop up on another. It was an expensive game of whack-a-mole.

"I spent nearly $3,000 in total before realising it was a big scam. The second they see you are sucker enough to pay for one it's up the next day on another," she said.[3]

Like Onree Norris and Breonna Taylor, Kim had inadvertently gotten onto the wrong list—lists that all of them would have done well to resign from, if only they could. The consequences in Kim's case were not life-threatening, but they were certainly threatening to her livelihood and reputation. The police debacles in the first two cases eventually brought words of regret from the officials involved, though the courts denied appeals for legal redress. But Kim's legal predicament was even more acute. The law afforded her no obvious means of preventing the cruelties that she was suffering from becoming chronic. The sources of demands for payment like those she experienced, according to the *Guardian* article, are shadowy companies that may claim protection under First Amendment guarantees of freedom of expression.

It's easy to put oneself in Kim's frame of mind as she struggled to cope with would-be blackmailers to avoid stigma for an offense for which she had been exonerated. She did not—and could not—challenge the fact that the mugshot in question was of her. But she must have asked herself, "What business of theirs is this mugshot?" "Where," she must have wondered, "did they get this information, and who told them that extorting payments for it was legal?" And finally, "Who *are* they? And why can't I put a stop to this?"

Kim may well have been targeted for this thinly disguised form of blackmail because the circumstances of her arrest—leaving a gay bar—rendered her especially vulnerable. There is a long history, after all, of blackmailing people over their gay identities. But in a world where every waking moment seems to leave its trace on a computerized feed of some kind, the possibilities for such misuse appear bound to rise.

Kim's case exemplifies a broad category of situations that cry out for protection. Let me call them *unilateral* personal-decision systems—private-sector systems that have no legal bases; that are not desired by their subjects; and whose consumers are not the subjects of the data files, but rather parties seeking to take some form of action toward those subjects. True, in the case of Mugshots.com, it's a little hard to specify who the consumers are. Victims of these enterprises pay the owners to desist, but one would hardly classify the targets of these operations as consumers. Unilateral systems, in this sense, exist wherever private parties have the wherewithal to create databases on subjects who have not sought to be monitored and who have no say in the decision-making taken in their cases.

Many such systems represent the privacy equivalent of weapons of mass destruction. Thus, my proposed Reform Ten:

> Anyone who finds herself or himself the subject of unwanted processing of personal data by a personal-decision system operating without legal basis, statutory authorization, or the subject's consent may insist that processing of those data cease, with immediate effect. Repeated failure to comply with subjects' demands along these lines will incur court judgments of compensatory and punitive damages for the operators of the systems.

A right like this would place obnoxious enterprises like those that assailed Kim on the defensive—challenging them to provide authentic legal bases for their work. It would also set limits to many, if not most, *data brokerages*—members of the vast American industry that compiles databases of personal information on "private" citizens and consumers, and that retails reports from these files for sale. Consumer credit reporting represents a borderline case for classification as a unilateral system, in that credit grantors in the United States sometimes seek consumers' consent to their capture and dissemination of at least some categories of the personal data they collect— though it's safe to say that most consumers lack the slightest idea of what they're being encouraged to agree to. But most data brokers ply their trade without the consent, and often without the knowledge, of those whose data they compile, process, and ultimately sell.

In other ways, the operations of data brokers parallel those of consumer credit reporting agencies—but with much more varied arrays of buyers and purposes for their reports. Some data brokers sell reports tailored primarily to planners for political campaigns, detailing matters likely to predict or to influence the subjects' political inclinations. These reports could include the mean income of the neighborhood where the subject lives; or the makes of cars registered to people on his or her street; or the forms of entertainment that he or she favors; or clubs and associations of which he or she is a member. Other data brokers generate reports more oriented to the subjects' consumption behavior—their tastes in recreation, for example, or the retailers' websites they visit. The buyers of such reports are apt to be advertising or marketing strategists, seeking outlets for their companies' products or services.[4]

Most Americans, I believe, would be astounded at the intimacy and scope of insight into the personal affairs of their fellow citizens marketed by one or more data brokers. Lists of names, addresses, and other identifying data can be purchased to order, based on criminal histories, family status, tastes in pornography, medical histories and current conditions, and countless other criteria—many of which hardly anyone wants to be identified as belonging to. Though they have their parallels in reports available from credit reporting agencies, these non-credit reports appear to be unregulated, at least by any federal law. Such companies fit the definitions of *unilateral* systems, in that they operate without legal bases or consent from the subjects of their reports. Often the subjects are quite unaware that such operations have their names in the first place. I advocate their proscription, for reasons given in chapter 2. Their abilities to track and disseminate intimate personal data on millions of people simply accrue powers too sweeping for any private party to wield safely. The costs of their going wrong outweigh any likely benefits to the broad public interest. Such powerful levers of human action need to be regarded with the caution and reserve accorded to *controlled substances*. Without legal bases and subject consent, I hold, they are simply too potentially destructive in their effects on the lives of their subjects to be acceptable. What data brokers have to sell are maps of our lives, including our vulnerabilities—both to parties known to us and to all the rest. For privacy's sake, it's better to make closed windows and drawn shades the default condition, and leave opening them to those who wish to grant their affirmative consent.

True, one can hardly level such arguments against *all* unilateral systems. A number of industries and other private parties have developed personal-decision systems that largely operate behind their subjects' backs—yet some of which may well enjoy significant public support. Consider personal-information systems associated with various forms of property insurance. Companies that insure personal property against theft and loss must deal with professional criminals who, if unchecked, purchase and insure supposedly expensive items like jewelry, which they then report as lost or stolen. It can be assumed that many such companies record and exchange personal information with other companies in an effort to avoid repeating their experience with such claimants. These are certainly unilateral systems. Their workings are not announced to the subjects of the files—and, of course, consent of the latter is not sought. But the industry would, no doubt, uphold these measures as elementary self-defense.

Much the same could be said for personal-decision systems maintained by physicians wishing to avoid accepting patients with histories of filing malpractice claims; or hotel owners seeking to decline business with travelers who stay but don't pay; or department stores and other retailers who seek to identify and refuse sales to consumers who they believe to be compulsive returners of expensive clothing or household appliances—or, indeed, tenant-screening services that seek to identify tenants responsible for major losses. I hardly contend that all these operations have equal claims on public support, and I certainly don't condone all the precepts or practices followed in their name. But it would be wrong to dismiss categorically the needs that all these unilateral systems claim to serve.

Defenders of institutions like these would recoil at any association of their organizations with outfits like those who sought to blackmail Kim—whose contributions to any larger social good appear to be zero, if not negative. Spokespeople for the other systems discussed above would presumably defend their activities as upholding the law, suppressing unethical or disruptive public behavior, or cutting costs that would otherwise be passed on to consumers. Those who serve the insurance industry, for example, will cast themselves as indispensable in preventing insurance fraud (which could raise the cost of insurance for honest consumers). Tenant-screening companies will claim to reduce losses experienced by landlords due to destructive or dishonest tenants, thereby (theoretically) reducing rental costs for other tenants. In short, different privately founded and maintained systems have different claims to operate, at least to some degree, in the public interest. No doubt, the parties on the receiving end of the surveillance in cases like these would jump at the chance to exclude themselves from such attentions. But granting such a "get out of jail free" card would presumably create incentives for insurance fraudsters, vandals of rental accommodations, or other opportunistic parties to pursue their destructive ends without check, at the expense of others.

A particular irony of situations like those discussed above is that establishing this "same" right would likely produce quite different results, depending on prevailing power relations in the industry where the surveillance takes place. Today, for example, America's consumer credit reporting system is relatively comprehensive, highly centralized, and thoroughly networked. By this I mean that any use of the system—from a credit card application in California to a request to rent an apartment in Omaha to an application for a homeowner's mortgage in New York—will likely leave its record in a credit file under the consumer's name, one that will be interrogated on further credit applications or credit inquiries from almost anywhere. Practically speaking, this means that the chances of escaping a "bad" credit record by moving elsewhere and starting again are poor.

Thus, in all sorts of ways, the American credit reporting system has made itself the "only game in town." Whether you're seeking to open a credit account or simply hoping to support an application for a new job, you will have trouble evading the effects of the contents of your credit file. Estimates of the proportion of American adults who have a credit file in their name put the figure at around 89 percent.[5] Unless you're a member of a monastic order or a long-term resident of a mental institution, you need credit, or at least a credit report—and for that a credit history. You are also likely, at some crucial junctures, to need to change jobs, rent or purchase a new residence, or obtain a character reference. Any executive in an organization seeking to make a crucial decision about anyone who cannot or will not produce a credit report will be tempted simply to drop consideration of the case. Even if Reform Ten were in place, those exercising a right to resign from credit reporting would likely end up simply excluded from many elements of what most of us would consider normal life—from the opportunity to rent a car to the ability to register in a hotel without making a major advance deposit.

By contrast, reporting systems with less complete coverage than that provided by America's credit reporting industry may be more adversely affected by exercise of a "right to resign." Imagine the situation in a city where no one company dominates tracking and reporting on tenants for the benefit of landlords, and where many landlords do not consistently report their experience with tenants to any company. Under these circumstances, when an applicant for rental housing gives a false previous address to a prospective new landlord, the latter may have no basis for knowing better. A report of "no record" in such a case may simply mean that information on the applicant is held in another company's files—but is inaccessible to the landlord who is inquiring. Here, a tenant's "right to resign"—in effect, a Fifth Amendment for renters—would offer a serious possibility of evading bad news from the past.

A Selective Right to Resign?

One could also consider establishing a "right to resign" on a case-by-case basis. What if subjects of credit reporting had the right simply to seal the records of specific credit accounts—for example, where there had been unresolvable disputes between creditor and consumer? America's current credit reporting system does not deal with such situations very well. They are likely to go down on the credit record as red flags against the consumer and, perhaps worse, lower the three-digit credit score that probably matters most in future applications for credit and other privileges. Today, consumers are permitted to enter a brief statement giving their own version of a disputed account for inclusion in subsequent reports to prospective credit grantors. But one doubts that buyers of these reports pay much attention to such statements. Statistical wisdom would probably counsel avoiding virtually *any* consumer with a history of disputes with credit grantors, regardless of the merits of the consumer's position in the matter.

But imagine a further change in the law, one that accorded every credit-using consumer the right to sequester from his or her credit record a finite number of "bad" accounts. What if, for example, any consumer could order the elimination of one "bad" account of every ten—or twenty, or fifty—account records. Imagine that such discretionary "censorship" of one's own credit record could be made without leaving any indication on the full record. A step like this might actually improve the power position of *all* credit customers by removing, or at least diminishing, the ability of any credit grantor to threaten any consumer by applying a black mark to his or her credit record. The credit grantor, in such cases, would never know whether the consumer had retained the option of suppressing from the record evidence of an unresolved dispute. The result could be the reduction of incidents where creditors unjustly threaten to punish consumers via credit reporting—and greater willingness on both sides to compromise disputed debts.

Regardless of what qualifications might be placed on such a right, objections from the affected industries are totally predictable. Permitting consumers to censor news of their failures to meet their financial obligations, industry spokespeople will insist, amounts simply to dulling the sword of surveillance justice. Retailers and other credit-accepting businesses will be no less indignant: after all, creating for consumers the ability to place any account "off the record" minimizes their incentive to comply with any disputed bill.

Algorithms and Justice

As techniques for compiling personal data become more formal—more fully translated into numbers, or binary "yes or no" responses—they become more subject to automated, computerized treatment, yielding directions for action that can be generated, and indeed implemented, by computer. As recently as the 1960s, most American credit reporting agencies universally maintained their subjects' credit histories in discursive, hard-copy form, with each of the subject's credit accounts occupying a line or so of typescript. Five decades later, virtually all this information is computerized and maintained in a series of succinct codes. Most significantly, the industry has distilled every consumer's desirability as a credit customer into a three-digit rating that can be used to administer cutoffs for credit applications (such as "accept all applications from consumers whose scores are 700 or above") or to modify terms of acceptance according to the strength of the applicant's rating (for example, interest charges for those below 750 are set at 12% per annum, while those applicants above 750 will be offered 10% interest).

These formal, quantitative methods of decision-making go by the names of *algorithms; artificial intelligence (AI);* or, among statisticians, *discriminant analyses*. Their purpose is to identify the quantitative ingredients of "successful" cases (from the standpoint of the organization conducting the analysis), so that decision makers can target those consumers whose numbers foretell success. In credit, of course, this involves the quest to initiate more accounts of the profitable sort and to reduce the proportion that yield losses—tasks readily delegated to computerized decision-making, based on formal analysis of data from credit applications in conjunction with applicants' credit scores.

Much the same logic now prevails in insurance marketing. Like credit, insurance involves selling promises of future obligations to the policyholder under specific future circumstances. As in credit, applications for insurance are increasingly processed using sophisticated analytical models aimed at predicting the likelihood of claims—synonymous with losses to the companies—under various future conditions. And as with credit, personal data provided in the application may provide bases for graduated offers of better or worse terms for coverage, such that the buyer pays more if deemed to be a high-risk customer. The applicant's credit score is today one of the most basic of these qualifying data.

If this account, thus far, strikes the reader as presenting insurance rate setting as a fact-based, dispassionate exercise in corporate strategizing, this is no accident. It is, in fact, the industry's version of its operations. Companies seeking acceptance for the pricing of their products, based on these quantifiable, formal analyses of risk, have no desire to revisit the historical precedents of their current operations. Before today's ascendance of quantification and formal statistical models came a period when the availability of housing, insurance, credit, and other consumer desirables was dominated by race-based discrimination that was all but universally understood in business circles, though not always publicly acknowledged. Both in the avowedly segregated Deep South and most of the rest of the country, Black people understood that they were unwelcome in most white neighborhoods, either as tenants or as homeowners. Lending institutions and the insurance industry closed ranks, making sales of homes to Black families in desirable white areas all but impossible. These practices went by the name of "redlining," after the use of maps (including some provided by federal agencies) that showed the neighborhoods to which Black people were to be restricted within red lines.[6]

Redlining received a major jolt with the passage of the Civil Rights Act of 1968, which made racial discrimination in the sale and rental of housing contrary to federal law. From that point, racial barriers to renting or owning a home were supposed to disappear. But movement in this direction has occurred only incrementally. According to a summary of recent research on 2020 U.S. Census data by sociologist John Logan of Brown University, "race, not income, is still the driving factor behind who lives where in the United States."[7]

Critics of these relatively enduring patterns of racial segregation point to what they characterize as cumulative disadvantages imposed on residents of minority neighborhoods. Costs of insurance are especially held to be greater for residents of Black neighborhoods than for otherwise similar white neighborhoods—interpretations actively disputed by the insurance industry.

Thus, a 2017 study sponsored by ProPublica and Consumer Reports provided a detailed look at costs and claims in California, Illinois, Missouri, and Texas, by residents in neighborhoods of different racial composition. The authors wrote: "Our findings document what consumer advocates have long suspected: Despite laws in almost every state banning discriminating rate-setting, some minority neighborhoods pay higher auto insurance premiums than do white areas with similar payouts on claims. This disparity may amount to a subtler form of redlining. . . . And, since minorities tend to lag behind whites in income, they may be hard-pressed to afford the higher premiums."[8]

The industry takes great exception: "This is a very, very serious charge being made on a very weak study," stated James Lynch, chief actuary of the Insurance Information Institute. "There is no unfair discrimination, intentional or unintentional," he concluded.[9]

Bitter, lengthy exchanges continue between the two sides. "I'm not surprised at the findings," stated Robert Hunter, director of insurance at the Consumer Federation of America. In the words of *New York Times* writer Ann Carrns, the latter organization "has conducted a series of studies raising questions about the fairness of using nondriving criteria, like education and occupation, in setting auto insurance rates. In 2015, the federation published a study finding that rates are much higher in minority ZIP codes."[10]

Against this view, industry defenders invoke the language of scientific analysis—as translated into the parlance of decision-making based on formal models. Insurance rates in minority neighborhoods are not high because the neighborhoods are inhabited by minority people, the arguments seem to go. Instead, they are high because the people who live there have distinctive habits of driving and of making claims, or perhaps because traffic patterns or sheer population density make accidents there more likely. In the words of industry defender Tony Cotto, "ratemaking is a prospective exercise—an effort to use the expertise of underwriters and actuaries, along with objective data statistically correlated to losses, to come up with the price at which the insurance company will take the risk."[11]

According to this view, to simplify only slightly, any contingency that someone might want to insure has a unique level of risk associated with it. The best any insurance company can do is to reflect the price of that risk in its charges. This should be done, according to Cotto, without giving weight to what he delicately refers to as attention to "characteristics . . . *sometimes limited by state law in the name of perceived fairness by policymakers*" (emphasis added).[12] The subtext here seems to be that such efforts to achieve "perceived fairness" (not authentic fairness, apparently) only interfere with the industry's single-minded determination to charge for each identifiable risk at its precise, objective cost. The author is not explicit about this, but it sounds as though he is referring to efforts by legislators or other rate setters to mandate equal—or, at least, more equal—insurance rates across racial lines.

At the heart of these debates is the concept of *risk* itself. Strictly speaking, for any given future event over any period of time, the probability of its occurrence is either 1 or 0. That is, all that we can be sure of in advance, about the likelihood of an accident's occurring to a specific car in a specific neighborhood over a one-year period, is that the event will either occur or not occur. Because analysts do not know which of these two eventualities will prove to be the case, they create models of how frequently similar incidents have occurred among populations of similar units. But the resulting probability models are not given by nature, like the speed of light or the relation between the diameter of a circle and its surface area. Instead, they are products of human imagination, reflecting the analyst's judgment of what constitutes an appropriate situation for comparison. The question is, on what basis are such systems of comparison constructed, and what considerations bear on their work?

There are many criteria according to which an analyst might rank risks associated with members of any group of applicants for insurance. One could measure the exact *height* of each applicant, charging higher rates to those whose heights were associated with greatest likelihood of claims, or greatest losses to claims actually paid. Or one could rank the complexity of retinal markings in the eyes of the applicant. Everyone's eyes have such markings, and they are not particularly difficult to rate in terms of complexity. Here, too, one could allocate higher insurance costs to those whose retinal markings were associated with the least profitable policies. Or one might choose any number of personal characteristics—physical, social, or behavioral—on which to collect personal information on all applicants for a given form of insurance and charge higher premiums to those whose statistics prove to be associated with the highest losses. Lower rates for applicants according to the number of clubs, associations, or other membership bodies to which they belong? Why not, if that statistic proved to be associated with the likelihood of an applicant's filing claims? Or one could develop what specialists would call a *multivariate analysis* of such likelihood, using a variety of measured characteristics for each applicant together in a single algorithm to yield the most precise possible prediction of expected profit or loss. If enough personal data were available, the predicting variables might range from retinal complexity to years of formal education to population density in the applicant's neighborhood.

But not all such efforts by companies to quantify the total risk chargeable to applicants for insurance coverage would be equally viable. The reason is that the public could not be convinced that height, retinal complexity, membership profile, or any number of other individual characteristics of motorists have anything necessarily to do with moral responsibility for generating insurance claims. Public acceptance of any such criterion for pricing insurable risks seems to depend on that criterion's entailing some element of perceived personal responsibility.[13] Drivers can thus be considered responsible for their personal records of accidents and citations. But retinal complexity as a criterion for setting the price of one's insurance would probably never pass muster in public opinion. The insurance industry has barely succeeded in recent decades in making its case for applicants' credit ratings as a determinant of "risk"—so that those with higher credit ratings can expect to pay less for car insurance. But even this proxy for "risk" appears to be a borderline case in public credibility—hence the prohibition against its use in a handful of states.

The closest that Tony Cotto comes to giving a comprehensive rationale for policy in these matters—policy on what should constitute reasonable subjects about which to seek personal information—is in the following chart:[14]

Considered	*Potentially Considered*	*Never Considered*
Location	Gender	Race
Claims History	Occupation/ Employment	Ethnicity

Age of Home or Vehicle	Education	National Origin
Type of Home or Vehicle	Marital Status	Religion
Driving Record (Auto)	Credit-Based Insurance Score	Income
Unfenced Swimming Pool (Home)	Criminal History	Literacy

The intended purpose of this chart, it seems, is to identify the forms of personal information that may prove publicly acceptable as bases for rate setting in homeowners' and car owners' insurance. The columns marked "Considered" and "Potentially Considered" seem to consist of variables that might carry at least a hint of personal responsibility for the making of insurance claims. The variables in the right-hand column, by contrast, would probably be politically unacceptable to much of the American public in the third decade of the new century—perhaps because they would imply continuity with the bad old days, when race represented an acknowledged principle of price discrimination.

Referring to the left and center columns of the chart, Cotto writes that these

> factors should be objective, actuarily sound, and have a credible, statistically significant correlation to expected losses and expenses. Because factors are correlative, the more information an insurer has and is able to use, the more accurately it will be able to assess the likelihood of loss.... While the exact characteristics considered are sometimes limited by state law in the name of perceived fairness by policymakers, the chart provides just a few examples of some of the hundreds of factors about a person seeking insurance and the item

being covered as well [as] examples of factors that companies do ***NOT*** consider [emphasis in original].[15]

Some of these supposedly acceptable factors (Gender, Marital Status, and Credit-Based Insurance Score, for example) sound suspiciously likely to raise issues of "perceived fairness" attributed to those troublesome figures identified as "policymakers." After all, hasn't the country already passed federal legislation outlawing price discrimination on the basis of gender—and haven't some states outlawed the use of credit scores in pricing insurance? And in any case, what do these variables have to do with the propensity to submit insurance claims? Would applicants' height or cholesterol count—or, indeed, the complexity of markings on their retinas—be valid variables for setting rates if a statistical association could be established with the likelihood of submitting claims?

As though to answer questions of this sort, Cotto remarks that "less information makes accurately assessing the likelihood of a loss more difficult," adding in a footnote: "The power of correlation is what makes any given factor valuable for enhancing accuracy." Factors do not "cause" a loss, he notes, "but they inform the likelihood of one."[16] Yet I would be willing to predict that Race, Ethnicity, National Origin, Religion, Income, and Literacy are all correlated with likelihood of loss through insurance claims. I cannot claim direct knowledge on these matters, but experience with normally potent variables like these in large data sets leads me to expect variables of this fundamental sort to be correlated, positively or negatively, with one another.

The reason why variables like race, religion, and ethnicity are not the subject of avowed intakes of personal data by insurance companies—that is, why the companies do not explicitly acknowledge basing prices for given forms of coverage on these variables—is that they amount to political dynamite. If the companies acknowledged that they were charging, say, Catholics, or low-income applicants, or persons of Lithuanian origin, more for any given insurance coverage—simply on the basis of documented statistical associations to this effect—public opinion would not take kindly to the news.

. . .

Lest the reader imagine that our discussion has wandered negligently far afield from matters of privacy, I beg forbearance. Throughout this book, I have argued for a view of present-day privacy controversies as driven by the need of organizations to manage processes of *discrimination* in dealing with large numbers of people. Whether the underlying discrimination processes are just or unjust, whether the ultimate aims are benign or coercive toward the subjects, these efforts to refine discrimination generate appetites for personal data as voracious as that of jet airliners for their fuel. Any evaluation of these uses of personal data must judge the intensity of the demand for information against the costs of appropriating it—costs both economic and practical, on one hand, and moral and political on the other.

If the ultimate criterion for acceptability of claims on personal information in the determination of insurance rates is predictive efficiency, then race, ethnicity, national origin, income, religion, and literacy ought logically to be given serious consideration as required personal-data intakes for any insurance application. The reason for not demanding these forms of personal information is that many people nowadays find them ethically disturbing—and for good reason. Since consumers have little or no control over the race or ethnicity ascribed to them, can it ever be fair to use such variables as factors in determining what they will pay for insurance, or for anything else? What Cotto seems to regard as unwarranted interference on the part of state regulators is actually an assertion of quite decent reluctance to penalize those whose only disqualification consists of such negative associations.

Origins of Lists That Shape Lives

One cannot review accounts of those whose lives have been snarled by surveillance gone awry without wondering how, exactly, the victims came to be selected as such. How does the transition occur between merely being eligible for some adverse action and becoming a real-life target? Consider this disturbing case, reported in the *New York Times* by journalist Kashmir Hill:

> On a Thursday afternoon in January [2020], Robert Julian-Borchak Williams was in his office at an automotive supply company when he got a call from the Detroit Police Department telling him to come to the station to be arrested. He thought at first that it was a prank.
>
> An hour later, when he pulled into his driveway in a quiet subdivision in Farmington Hills, Michigan, a police car pulled up behind, blocking him in. Two officers got out and handcuffed Mr. Williams on his front lawn in front of his wife and two young daughters, who were distraught. The police wouldn't say why he was being arrested, only showing him a piece of paper with his photo and the words "felony arrest" and "larceny."
>
> His wife, Melissa Williams, asked where he was being taken. "Google it," she recalls an officer replying.

Williams spent the night in a detention center, where the authorities took his mugshot, fingerprints, and a DNA sample. The next day, he faced interrogation by two detectives. One produced a photograph taken in an upscale Detroit shop, purportedly showing a shoplifter at work. "Is this you?" one detective asked. "No, this is not me," Williams replied, holding the image produced by the detective next to his face. "You think all Black men look alike?"

Williams had been identified as the shoplifter, on the basis of facial recognition software, which had linked footage taken at the crime scene to his name. Confronted by the obvious discrepancy between the photo of the shoplifter and Williams's living presence, the detectives drew back. Williams recollected that "one detective, seeming chagrined," said to his partner, 'I guess the computer got it wrong.'"

> Mr. Williams asked if he was free to go. "Unfortunately not," one detective said.
>
> Mr. Williams was kept in custody until that evening . . . and released on a $1,000 personal bond. He waited outside in the rain for 30 minutes until his wife could pick him up. When he got home at 10 p.m., his five-year-old daughter was still awake. She said she was waiting for him because he had said, while being arrested, that he'd be right back.
>
> She has since taken to playing "cops and robbers" and accuses her father of stealing things, insisting on "locking him up" in the living room.[17]

Ultimately, these events seem to have triggered critical reactions even among Detroit law enforcement authorities themselves. In response to this *New York Times* report, the Wayne County prosecutor's office announced that Williams could have the case and his fingerprint data expunged. "We apologize," the prosecutor, Kym L. Worthy, said in a statement, adding, "This does not in any way make up for the hours Mr. Williams spent in jail." Given that Williams had an alibi for the crime that could have been checked before he was taken into custody, the brutal approach to his case represents massive overreach. And it is difficult to imagine that this would have occurred in the same way and gone uncorrected for as long as it did, had Mr. Williams not been Black. Clare Garvie, a senior associate at Georgetown University's Center on Privacy and Technology, commented, "I strongly suspect this is not the first case to misidentify someone to arrest them for a crime they didn't commit. This is just the first time we know about it."[18]

Seen in proper perspective—as an arrest based on flimsy evidence—these events must trigger incredulity in almost anyone. But what most demands attention here is what one might call the divergent *prognoses* that emerge from this horrific event—that is, what conclusions to draw from it and events like it, and what steps to take in response. These questions are by no means unique to this case, but in fact demand concern in the context of a wide array of technologically abetted acts of overreach. At one end of this continuum are those who define events like those experienced by Williams as abuses or errors—malfunctions to be corrected by more rigorous practice. At the other end of the spectrum are those bracketing such outcomes as endemic or systemic—something that everyone must simply become inured to, as the technologies come into wider use.

The first group appears to include Georgetown's Garvie. "There are mediocre algorithms and there are good ones," she declared, "and law enforcement should only buy the good ones."[19] Nevertheless, she seems to believe, mistakes are remediable. "Mediocre" algorithms, apparently, stand to be improved. The second form of thinking would regard outcomes like the arrest of Robert Junian-Borchak Williams as endemic in the use of the technologies—with "technologies" understood broadly, as including both physical tools like video recorders and the knowledge and understandings necessary to put such tools to work. If the users of the technology are not prepared to live with the occasional "false positive" of this sort, decision makers should abandon these tools altogether. Apparently inspired by such thinking, activists have succeeded in pushing municipal governments to ban, or severely limit, the use of facial recognition technology in at least a dozen American cities, including Somerville, Massachusetts, San Francisco, and Portland, Oregon.[20]

At about the same time that facial recognition was first drawing attention in the general media, some remarkable research on the reliability of these techniques was shaking up academic discussions of the subject. Some young computer scientists published the results of a series of empirical investigations demonstrating considerable unevenness in the ability of various algorithms to identify people—with inabilities skewed distinctly according to subjects' skin color and gender. Available algorithms designed to associate names with faces succeeded far less often, the darker the skin tone of the subject. And the worst performance, in this respect, appeared to be biased against women, with gender being a major source of error in its own right. According to researchers Joy Buolamwini and Timnit Gebru, "darker females have the highest error rates for all gender classifiers," ranging from 20.8 percent to 34.7 percent, while "lighter males are the best classified group," with 0.0 percent and 0.3 percent error rates, respectively. "When examining the gap in lighter and darker skin classification, we see that even though darker females are most impacted, darker males are still more misclassified than lighter males."[21]

In some rare manifestation of ironic justice, the authors of this now famous study are both women of African descent. Both in the published paper and subsequent commentary, Buolamwini and Gebru have called attention to the human costs of reliance on these error-prone technologies. The shock waves emanating from their findings have nowhere been greater than in the community of computer professionals. One result has been Buolamwini's founding of the Algorithmic Justice League, an activist group dedicated to raising "awareness of the social implications of artificial intelligence through art and research."[22]

In an attention-grabbing gambit, the American Civil Liberties Union dramatized the stakes involved by submitting photos of all the current U.S. representatives and senators to Rekognition, a facial identification technology developed by Amazon, asking it to compare the images with some twenty-five thousand publicly available mugshots. The results: twenty-eight members of Congress were misidentified as belonging in the mugshot sample—or a 5 percent error rate. "The test disproportionally misidentified African-American and Latino members of Congress," according to Natasha Singer, writing in the *New York Times*.[23]

An Amazon spokesperson insisted that during its test the ACLU had used the company's face-matching technology differently from the practices the company recommended for law enforcement. Police departments, she said, "do not typically use the software to make fully autonomous decisions about people's identities," and "in real-world scenarios, Amazon Rekognition is almost exclusively used to help narrow the field and allow humans to expeditiously review and consider options using their judgment."[24]

Credit Surveillance in the United States, Australia, and France

By now, I hope readers can agree that, in a world shaped profoundly by the workings of personal-decision systems, personal information plays a role much like that of coal during the industrial revolution. Without steady flows of just the right kind of personal data, organizations ranging from law enforcement agencies to insurance companies to medical care providers and customs and immigration authorities would risk operational paralysis.

But this generalization hardly warrants the conclusion that all resulting demands for personal data deserve our support. As examples in this chapter have shown, some appetites for more personal information simply don't justify the cost. The many possible systems for creating, compiling, and using personal information have widely varying impacts on privacy—forcing us to turn our critical attention to the surveillance *institutions* implementing the need for data, and to judge which of these appear to do their work at the most acceptable cost to other key values.

In such judgments, history matters. Different traditions of gathering and using personal data, once begun, can persist with great effect over generations. In the United States, by the end of the nineteenth century, entrepreneurs in the largest cities had begun to sell detailed listings of many local consumers, marking the beginnings of this country's credit reporting industry.[25] These publications detailed the reputations of each as to his or her reliability in paying their bills as well as their general character—the two apparently being all but congruent with one another in local reputation systems of the time.

These listings were necessarily static, in that they could not easily be updated until a new edition of the community-rating book appeared. And they missed details on consumers' lives that many creditors now consider essential: Was the consumer in danger of becoming unemployed? Or getting divorced? Or engaged in a new business venture, involving new demands on his or her financial resources? Credit *bureaus*—businesses dedicated to retailing intelligence on consumers' use of credit—emerged early in the twentieth century to compile current intelligence from credit grantors and incorporate more and more such contextual information in their files and reports. By the mid-twentieth century, the credit reporting industry consisted of thousands of such independent bureaus dispersed in medium to large towns and cities, concentrating mainly on tracking local credit use by local consumers. Outsiders could obtain a report on a local consumer by ordering it from the appropriate local bureau, but only if the buyer of the report knew where that consumer's records were held.

Beginning late in the 1960s, the industry underwent a sweeping transformation, eventually evolving from what had been a highly dispersed, capital-poor industry of thousands of locally oriented businesses into three nationally centralized giants—Trans-Union, Equifax, and Experian. This transition coincided with the computerization of the industry. From the treatment of individual consumers' records to the transmission of reports, electronic storage replaced file cabinets and paper within little more than twenty years—vastly increasing the speed with which credit data were reported and recorded and permitting the computerized decision processes described above.

Since roughly the 1990s, credit records have consisted of two forms of personal data. First are detailed listings of past and present credit accounts, showing promptness of payments and any delinquencies, along with personal data on the subject's current and past employers and information drawn from public records on matters like bankruptcies, tax liens, foreclosures, and repossessions. Second is what has come to be perhaps the most crucial single scrap of surveillance information in the life of any American consumer: the three-digit credit score, based on the ensemble of data available to the agency and supposedly epitomizing the consumer's desirability as a credit customer. Today, of course, with the computerized operations of nearly all American credit reporting companies, requests for credit reports and responses take place in real time.

Today, American credit reporting is nationwide in scale. The three industry giants aim to include in their reports information on the same consumer drawn from any part of the country—from a local retail account with an L.L.Bean outlet in Maine to multistate use of an American Express card or an airline credit card. More important still, drawing a credit report on any American consumer normally triggers creation of a record of that event in at least one of the three major credit reporting companies. That record is quickly shared, so that an updated credit report on the given consumer comes to include all current and recent credit accounts—giving something close to a comprehensive portrait of the consumer's current and recent obligations. To this ever-flowing stream of data from companies holding the consumer's accounts, American bureaus are likely to add personal data from public record sources—property liens, bankruptcy filings, and data on change of employment or marital status.

"Positive reporting" is the winsome name that the American credit reporting industry has given to this system of recording details of all, or nearly all, of a consumer's active credit accounts. It sounds, well, *positive*—as though it stemmed from a desire to recognize virtuous consumer behavior, rather than simply reporting bad credit news about accounts that have gone sour. But the practice has what one suspects is a more important virtue for the industry. It affords an intensity of surveillance that enables shrewd credit managers (or their algorithms) to spot instances where a given consumer, without yet slipping into delinquency on any account, has become "loaded up" (in the jargon of the industry) to the point where delinquency may not be far in the future. Having spotted a signal like this, the response of American credit grantors is typically either to seek a raise in the interest rate on the account in question or a reduction in the amount of credit available. In situations like these, the last thing any credit grantor wants is to be the last business holding big debts from a consumer who can no longer pay.

One can think of all these American institutions and practices as responses to pervasive "needs" for personal data. Yet similar needs have been fulfilled quite differently in other environments. Australia and France are both flourishing consumer economies, for example, yet until recently they have followed distinctly less invasive paths toward satisfying the informational needs of credit grantors. France still has no consumer credit reporting industry in the American sense. With one important exception, French credit grantors reviewing applications for personal loans, mortgages, debit cards, and the like long had no independent way of knowing what other credit obligations an applicant had outstanding. Still today, banks, retailers, and other credit grantors must assemble their own dossiers on each applicant, largely on the basis of information furnished by that applicant—in the form of pay stubs, for example, or receipts for major monthly expenses, or bank account statements. Those reviewing the applications may make inquiries to parties noted by the applicant as holders of other credit accounts. But, generally speaking, applicants have better possibilities for concealing details of their financial affairs than they would in the more aggressive surveillance environment of the United States.

Until recently, there has been just one *outside* source of information that French credit grantors could bring to bear on their decisions, regardless of the wishes of the applicant. This is a report from the FICP (Fichier national des incidents de remboursement de crédits aux particuliers). This database, maintained by France's central bank for roughly the past thirty years, lists consumers with accounts in arrears. This listing, to which all creditors are required to report their delinquent accounts, serves as a blacklist. Any applicant listed here will certainly see his or her chances of being approved for further credit much diminished. Lenders who do extend credit to those on the list can expect difficulty under French law in enforcing their debts. This practice significantly alters the balance of advantage between credit seekers and credit grantors, in that the costs of "bad" allocations of credit—those that end in nonpayment—are apt to be shared between the two sides.

Australia does have its own credit reporting industry. But there, arrangements resembling the classic French system prevailed until the enactment of legislation adopted in 2014. Before that, details of currently active credit accounts rarely found their way into Australian credit reports. The most consequential data conveyed there were the numbers of *inquiries* on the consumer recently registered from prospective credit grantors. Any spike in such inquiries would normally signal to prospective credit grantors that the consumer was experiencing an urgent need for new resources—never auspicious for granting further credit. Australia also maintained a national registry, much like France's, of credit accounts that had become legally delinquent. So long as this regime prevailed, a "good" credit report in Australia was one that held little detail—with few recent inquiries from other creditors and no report of current delinquency. As in France, credit grantors evaluating requests for new credit were expected to make their own inquiries about the applicant's current account. Also as in France, Australian courts were less sympathetic to lenders that had apparently ignored clear danger signals from standard sources of information on the applicant.

Note the differences between the American and the classic French and Australian systems, in terms of the demands they impose for detailed personal data. The latter two countries' approaches relied heavily on centralized national registries to document established accounts currently in delinquent status—that is, evidence of demonstrable, current trouble of the sort that prospective credit grantors seek to avoid. The more ambitious American model, by contrast, has long aimed to track consumers' current financial lives so completely as to foretell when they are likely to find themselves unable to meet credit obligations that may, or may not, develop in the future.

If the two models described medical examinations, rather than credit checking, the American approach would be classified as far more invasive. The principle that less news is good news does not require those making credit decisions to track every single financial obligation and strand of credit use that might, eventually, bear on the applicant's ability to meet credit obligations. Under this regime, credit checking represents a more blunt instrument for assessing the financial situations of consumers—simply because some risky forms of credit use (for example, purchases that strain the borrower's budget without quite breaking it) and some instances of notably prudent personal credit management may not be apparent on available records.

The winner, in these two countries, is privacy. Or at least it was.

Unfortunately, the virtues of these two less invasive approaches have been losing ground to aggressive proselytizing on behalf of the American model. In the past fifteen years, both France and Australia have come to permit reliance—without the applicant's permission—on information on current credit accounts that are *not* in arrears in evaluating credit applications. In France, lending institutions considering credit applications may now obtain account data on the applicant from other institutions under the same corporate ownership as the lender. This opens many possibilities for disclosure of credit accounts that the applicant may prefer not to list on his or her application. And in Australia, since 2014, both buyers and producers of credit reports on their buyers now have access to what the Australian industry have elected to call "comprehensive" credit information—roughly the equivalent of America's "positive" credit reporting.[26] This step has substantially undone the "no news is good news" system that had prevailed for decades.

France's national privacy-protection watchdog, the CNIL (Commission nationale de l'informatique et des libertés), actively resisted the changes, taking a public stand in defense of the classic system. As in Australia, it appears that the success of the American model in reshaping credit reporting in France reflects the political weight of that country's banking industry.

Conclusions

It is obviously impossible for citizens and consumers to imagine a right to "resign" from personal-decision systems, if they have no more than a shadowy understanding of what those systems might consist of. Since the databases used to set their rates for credit and insurance are beyond many consumers' personal awareness, we can only imagine what their response might be to an opportunity to resign from those systems. Onree Norris and Breonna Taylor, the targets of potentially deadly no-knock domestic invasions, ought to have had the chance to contest the records portraying them as part of the drug trade before the authorities came crashing through their front doors. But they were probably also unaware of the very existence of the personal-decision systems that were targeting them. We don't have to wonder what Robert Junian-Borchak Williams would have thought about the information that led to his arrest in front of his assembled family, since he has been articulate on that subject.

From the other side of the individual/institutional divide, by contrast, there are always well-organized interests—corporate and governmental—ready to promote a more closely monitored life for all. In the private sector, incentive for such intensified surveillance often seems to stem from the desire to exclude unprofitable prospective customers while recruiting those promising the highest potential returns. In the public sector, the parallel to this impulse seems to be willingness to accept even proactive measures against supposed sources of narcotics or other activities deemed especially noxious, as a reasonable cost of this country's massively ineffective drug control establishment. The idea that coming down very hard on those at the bottom is somehow necessary for a decent life for everyone else sometimes seems to be an American specialty.

But this hard-hearted outlook does not represent the only face of American surveillance politics. In 1974, on a wave of demands from Americans for justice along racial and gender lines, Congress passed the Equal Credit Opportunity Act that effectively barred discrimination based on race and gender in the allocation of various forms of credit. This outcome, let us note, represented an embrace of what the insurance industry's public statement, as quoted above, delicately termed "perceived fairness." The legislators of 1974 were saying, implicitly, that access to credit for minority and female applicants should be equal, *even* if statistical analysis showed different levels of risk between them and white males.

At least, I hope that was what the representatives and senators were thinking, because it seems to me the most auspicious attitude for encouraging engagement in the system by parties with histories of less-than-satisfactory relations with a consumer economy dominated by white males—the best social therapy, if you like, to redress the effects of a troubled, unequal past. One can think of all sorts of plausible explanations for why women and racial minorities might have records of credit use that are inferior to those of typical white male customers, by the standards of the industry. Suppose, as many claim, that both females and Black people had been targets of overt discrimination in credit markets for decades—and that alienation from credit-granting institutions or unfamiliarity with credit-using strategies had led to poor credit performance by members of the two groups. Would such a finding represent justification for applying stricter decision rules against the two groups into the indefinite future? Or would a more reasonable response be to equalize the opportunities for these groups—in hopes that the longer-term result would be to move the system toward a new equilibrium where race and gender have no association with payment patterns?

Notes

1. Nicholas Bogel-Burroughs, "Ex-Detective Admits Misleading Judge Who Approved Breonna Taylor Raid," *New York Times*, Aug. 24, 2022, A-14.

2. Evan Hendricks, "Privacy Times," *Privacy Times* 26, no. 8 (Apr. 2006).

3. Olivia Solon, "Haunted by a Mugshot: How Predatory Websites Exploit the Shame of Arrest," *The Guardian*, June 12, 2018.

4. Epsilon, "Consumer Information," https://www.epsilon.com/us/consumer-information.

5. Bureau of Consumer Financial Protection, "The Consumer Credit Card Market [2021 Consumer Credit Card Market Report]," Sept. 2021, https://files.consumerfinance.gov/f/documents/cfpb_consumer-credit-card-market-report_2021.pdf.

6. Julia Angwin, Jeff Larson, Lauren Kirchner, and Surya Mattu, "Minority Neighborhoods Pay Higher Car Insurance Premiums Than White Areas with the Same Risk," *ProPublica*, Apr. 5, 2017.

7. "Making Sense of the Census: In 2021, America Is Less Racially Segregated, but No Less Unequal," *News from Brown* (Brown University), Oct. 4, 2021, https://www.brown.edu/news/2021-10-04/.

8. Angwin et al., "Minority Neighborhoods Pay Higher Car Insurance Premiums."

9. Ann Carrns, "Study Finds Car Insurers Raise Rates in Minority Neighborhoods, *New York Times*, Apr. 5, 2017.

10. Ibid.

11. Tony Cotto, "NAMIC Issue Analysis: Why Your Insurance Costs What It Does," *National Association of Mutual Insurance Companies*, Jan. 2021, 5.

12. Ibid.

13. For a penetrating discussion of the interactions of causal understanding and moral categorization in the public acceptance of insurance pricing, see Barbara Kiviat, "The Moral Limits of Predictive Practices: The Case of Credit-Based Insurance Scores," *American Sociological Review* 84 (2019): 1137–58.

14. Ibid.

15. Ibid.

16. Ibid.

17. Kashmir Hill, "Wrongfully Accused by an Algorithm," *New York Times*, June 24, 2020.

18. Ibid.

19. Ibid.

20. Shannon Flynn, "13 Cities Where Police Are Banned from Using Facial Recognition Tech," *Innovation and Tech Today*, Nov. 18, 2020, https://innotechtoday.com/13-cities-where-police-are-banned-from-using-facial-recognition-tech/.

21. Joy Buolamwini and Timnit Gebru, "Gender Shades: Intersectional Accuracy Disparities in Commercial Gender Classification," *Proceedings of Machine Learning Research* 81 (2018): 1–15, 11.

22. Algorithmic Justice League, https://www.ajl.org/learn-more.

23. Natasha Singer, "Amazon's Facial Recognition Wrongly Identifies 28 Lawmakers, A.C.L.U. Says," *New York Times*, July 26, 2018.

24. Ibid.

25. Josh Lauer, "Coming to Terms with Credit," in *Creditworthy: A History of Consumer Surveillance and Financial Identity in America* (New York: Columbia University Press, 2017).

26. Office of the Australian Information Commissioner, 2021 *Independent Review of the Privacy (Credit Reporting) Code,* September 2022, 18–19.

6 Create a Property Right over Commercialization of Data on Oneself

When historians set out to chronicle the defining economic transformations of our times, the exploding value of personal information will surely loom large. The salient corporate success stories of the late twentieth and early twenty-first centuries have many and varied origins. But "GAFAM"—Google, Amazon, Facebook, Apple, and Microsoft—all owe their ascent largely to mastery of data on those they deal with. These data range from users' buying habits and web-surfing interests to their political reflexes, family values, and social network connections. Resourcefully captured and harvested, sliced and diced, combined with other data, and analyzed so as to wring out the subtlest personal insights, such information provides decisive advantage to those who know how to employ it. Nor is it only industry giants that owe their success to these measures. Organizations of all sizes have learned that knowing *just* the right things about those they are dealing with can enable them to foretell what those people will do, or want, or be open to trying—perhaps even before the latter know these things themselves.

It was not always like this in American business. Consider the beginnings of the New York and Mississippi Valley Printing Telegraph Company, in 1851. The founders of this early information-industry startup—now known as Western Union—seem to have felt no need to scrutinize the consumption habits, moods, or personal associations of would-be customers before offering their services. Nor did Henry Ford show much interest in the social network connections or recreational habits of potential buyers of his Model T. By contrast, today's corporate strategists, confronting a public with significant disposable income and many avenues for spending it, feel called upon to know their customers' lives in strategic depth.

State agencies are not much different. Today, countless government offices track the fine details of the lives of the governed. Their aims in doing so include assessing and collecting taxes; enforcing compliance with environmental protection laws and regulations; licensing vehicles and drivers; and administering social security, health care delivery, zoning, and many other government functions. Much as in the private sector, public administration often requires close calibration of government action to the fine detail of the lives of the governed. More than in the private sector, the biggest government personal-decision systems—those maintained by agencies like the IRS, Social Security, and state motor vehicle departments—approach total coverage of the adult population. And beyond such relatively open government systems, remember the shadowy institutions of domestic surveillance maintained by agencies like the Department of Homeland Security, the NSA, the FBI, and the many state and regional "Fusion Centers." Accordingly, government agencies are among the major purchasers of Americans' personal data from the country's many data brokerage companies.[1]

The seemingly never-ending quest by large organizations to know more and more about those they deal with leads to the pervasive exchange of personal information among organizations—that is, to *markets* in personal data. Recall the "scraping" operations revealed in a *Wall Street Journal* article cited in chapter 3. Investigative reporters Julia Angwin and Steve Stecklow revealed how aggressive researchers captured even patients' intimate reports of their symptoms and medications for sale on the internet. Sometimes government agencies are the purchasers in these transactions, and sometimes they are the sellers.

Table 2 shows the prices offered for various personal data in some of the most active US markets for such information.[2] The origins of information on consumer credit reporting are different, mainly because the volume of reports sold is so great—with an estimated 240 million reports sold in 2020—and the number of organizations involved so large.[3] Virtually all credit reports on consumers throughout the United States originate from three giant companies: Experian, Equifax, and TransUnion. The prices for these reports, however, vary enormously from buyer to buyer, according to the volume of reports purchased by any one buyer and the level of detail of the report.

Note that, whereas the prices of personal data shown in this table are relatively small on a per name basis, the aggregate of these activities makes up a healthy industry. Indeed, companies in positions to capture and resell information on matters like the locations of cell phone users find this activity so lucrative that they have been willing to risk the wrath of regulatory agencies in order to help themselves to this opportunity.

At the time of this writing, the telecommunications industry finds itself in deep trouble with its federal regulators for pursuing the "finders-keepers" practices toward personal information so widely embraced in this country. A *Wall Street Journal* article from February 2020 tells of industry-wide efforts to evade obligations to keep the locations of cell phone users secret. Internet service providers are of course privy to this information, which holds enormous value for retailers seeking to advertise their services to cell phone–using

TABLE 2 Prices Advertised for Personal Information on Various Categories of Americans

Personal information	Name of company	Description (from company website)	Basic service	Basic price
Criminals and sex offenders	BackgroundChecks.com (nonsexual crimes)	"[S]ourced from a name and address history based on the subject's social security number. The names searched for include the primary name and other names found in the name and address history for your subject, such as maiden names, previously married names, middle names used as first names, nicknames and other aliases. Our database contains over 650 million American criminal convictions."	Criminal database search, alias verification, and instant non-hit results	$29.95 per inquiry
	BackgroundChecks.com (sex offenders)	"Backgroundchecks.com will search our own database, that includes a state's specific sex offender registry data and available tribal registries in that state, based on the applicant's name and date of birth. We have access to comprehensive sex offender registry data, with information from 49 states, plus Washington DC, Guam, Puerto Rico and various tribal sex offender registries, with photos in all. Details in the report include: identifiers, registered addresses, aliases, case numbers, charge(s), conviction details, and period of incarceration."	Sex offender registry search	$6.95 per inquiry

	Contemporary Information Corp. (cicreports.com)	"CIC's accurate and timely public records information is designed for consumer reporting agencies, resident screening providers, Software as a Service (SaaS) platforms, research institutes, colleges, universities, and online information services."	SSN trace, national criminal database scan, and manual county court search	$25.00 per inquiry
	Criminal Watchdog, Inc. (criminalwatchdog.com)	"CriminalWatchDog, Inc. is the foremost authority for accurate, comprehensive and innovative criminal background check and pre-employment screening solutions utilized by the Fortune 500 and small businesses throughout the United States. With nearly 465 million criminal records on file, from over 50 states."	Felony, misdemeanor, sex offender, inmate, probation, and "other state and county criminal offense records nationwide"	$18.95 per inquiry
	Accurate (accurate.com)	"By optimizing every facet of the client and candidate experience, we deliver the background checks you've always wanted."	SSN trace, address history, current county criminal search, national criminal check with county verification, and national sex offender registry	$29.95 per inquiry
Financial and property	Property Shark (propertyshark.com)	"[A] real estate website that provides in-depth data for approximately 90 million properties in New York City, Los Angeles, San Francisco Bay Area, and other major US markets. Named one of the Top 50 Best Websites of 2009 by TIME.com, the website caters mainly to real estate professionals, investors, and home buyers. While a free basic membership option is available, a paid subscription is required in order to access some features and tools."	Owner names and mailing addresses, defaulted mortgages potentially headed for foreclosure, construction liens, sidewalk liens, tax liens, and condo common charges liens	$79.95 per inquiry

(continued)

TABLE 2 (continued)

Personal information	Name of company	Description (from company website)	Basic service	Basic price
	A Good Employee (agoodemployee.com)	"We Make Background Screening Simple. A Good Employee is a one-stop resource for your employee background screening needs. We help companies of all sizes make smart hiring decisions. We recommend a financial background report for job positions that require managing your company's finances or sensitive financial data."	Collections, tradelines, inquiries, past-due accounts, bankruptcies, tax liens, and civil judgments	$19.95 per inquiry
LGBTQ+ Identity	Paramount Direct Marketing (paramountdirectmarketing.com)	"List of all people in database with 'Alternate Lifestyles': These households have reported that they engage in an alternative lifestyle. They are part of the Gay, Lesbian, Bi or Transgender community. They have visited various gay websites, made purchases, inquired about clothing, travel and more."	List of 600,000 names of members of the LGBTQ+ community	$110.00
Health	Paramount Direct Marketing	"Paramount Direct Marketing is an advertising agency that specializes in targeting new customers via direct mail lists, telemarketing lists, email marketing, and display advertising."	List of 7 million names of people suffering from mental health problems	$90.00
	Paramount Direct Marketing		List of 300,000 names of people suffering from ADD/ADHD	$180.00

Paramount Direct Marketing	List of 3.5 million names of smokers planning to quit	$125.00
Paramount Direct Marketing	List of 1 million names of e-cigarette smokers	$125.00
Exact Data (exactdata.com)	"Exact Data sources consumer data from national database with approximately 210 million names, postal addresses, and telephone numbers, with approximately 700 selects. The database is compared to the USPS National Change of Address file every 60-days, and updated as necessary."	List of 15,005 names and addresses of consumers suffering from Alzheimer's disease	$0.44 per individual
Exact Data	List of 238,189 names and addresses of consumers suffering from heart disease	$0.08 per individual

Note: Table 2 shows prices offered for various forms of personal information on Americans as of January 2021. The intended buyers of these reports include would-be employers (concerned to avoid hiring employees with criminal records); owners of rental properties (seeking to avoid troublesome tenants); and marketers seeking to identify potential buyers of products and services ranging from gay-friendly vacations to cures for smoking. Not included here are activities of the huge consumer credit reporting industry, which sold an estimated 240 million reports in 2020. Virtually all credit reports on consumers throughout the United States originate from three giant companies: Experian, Equifax, and TransUnion. The prices for these reports vary enormously from buyer to buyer, with the shortest, least expensive reports costing as little as ten dollars to highly detailed reported for transactions like mortgages that cost well more than one hundred dollars.

customers, who may, at the same time, have strong reasons of their own for not wanting their locations made available.

The carriers, including AT&T, Sprint, T-Mobile US, and Verizon, faced fines of more than $200 million for regularly tipping the locations of their subscribers to intermediaries, who then retailed the information to what are described as "hundreds of businesses." Some privacy advocates criticized the FCC action as overdue.[4] Laura Moy, associate director of the Center on Privacy and Technology at Georgetown Law School, pointed out that consumers have no choice but to make highly personal information available to their service providers. Yet the latter appear to have widely violated that trust, she insisted, and the FCC has been slow to act.[5]

In short, it is worth a lot of money to many companies—and to government agencies as well—to track the locations of Americans and other details of their lives. This information enables retailers to bombard travelers with ads for goods and services that pop up on their screens just when they happen to be nearing the businesses selling them. As earlier chapters have shown, knowing when specific customers are passing specific retail establishments effectively opens a gold mine for retailers willing to act quickly.

Personal Information Property

But note something remarkable about these lucrative data flows. The data being bought and sold here originate in—and gain their interest from—the lives of *ordinary* Americans. The effects of these transactions on those lives are often decisive. *Celebrities* can legally claim compensation for commercial uses made of their name and fame—interests that can live on after the celebrity has left the scene. According to one recent calculation, Elvis Presley was one of the "top four dead celebrities," earning more than $27 million in royalties and other fees in 2016.[6]

Most of us, living or dead, cannot expect to derive such benefits from commerce in the details of our lives. But what if we all *owned* rights over the *commercial exploitation* of information on ourselves—that is, the use of our names, our images, and the facts of our lives to facilitate profit-making or fund-raising efforts and services, through such activities as advertising, consumer credit and insurance reporting, electioneering, and the like? This new right over our own data would resemble other property rights—water rights, mineral rights, air rights—all of them associated with real property, but capable of being sold or leased separately. Surely a right like this would provide the basis for a strong defense of privacy, even while permitting those so inclined to allow commercialization of data on themselves.

This new property right would prevail regardless of where or how the data in question originated. Personal information ranging from our salaries, to our medical histories, to reports of our movements and personal associates would then have blanket protection against appropriation—though solely for commercial purposes. The default condition, where the subject of filed data simply did nothing, would be to block all commercial use, broadly understood. Such use would include the following:

- Use of the subject's image or the details of his or her life in advertising
- Manipulation of the subject's data to facilitate commercial or money-raising activities—for example, consumer credit reporting, or the sale of lists of prospects for charitable appeals or of voters as potential targets for appeals by candidates and causes
- Records of subjects' interactions with online services that might serve as bases for fund-raising or moneymaking appeals of a

commercial or political nature—as in the models of internet users' susceptibilities or motivations developed through analysis of their interactions with systems like Google or Facebook
- Any other attempts to derive monetary value from the unique or distinctive details of people's lives or actions

Surely the resulting tension between owners of our personal data—that is, each of us—and those seeking to commercialize such data would raise the salience and viability of privacy interests. Simply by doing nothing, anyone and everyone could block all forms of commercialization of his or her data—from appropriation of cell phone usage records for targeting ads; to sale or trade of mailing lists from voluntary associations or political organizations; to paid dissemination of internet users' psychological susceptibilities to political messages, as in Cambridge Analytica's attempted intervention in the 2016 presidential campaign.

Strong privacy legislation in the European Union already blocks most of what Europeans call "secondary release" of personal data. From banks and credit card companies to mail-order houses, periodicals, and charities, the law forbids the release, without permission from the subject, of personal information compiled for one purpose to other parties for new purposes. But Americans could create an even more powerful and responsive set of privacy controls by establishing the property rights put forward here.

Thus, Reform Eleven:

> No personal information held in a personal-decision system may be sold, rented, "shared," or traded for value for any commercial purpose, except as required by law; by an otherwise valid contract with the subject of the data; or with the latter's freely given consent. Such consent may legally be predicated on payment of some form of royalties to the subject of the data.

This right would apply to any personal information held in the form required for use in a personal-decision system—whether furnished by the individual in question or by someone else. It would apply to data about the individual "mined" by third parties—for example, from court records, records of businesses and nonprofit organizations, records of government bureaucracies, and data collected through use of "cookies" implanted without consent on consumers' computers. It would include sales of information compiled for purposes of credit screening or insurance vetting, "background checking" on employment applicants, placement of online ads via social media such as Facebook or Google, or targeting of voters in election campaigns. Consent from the individual concerned would still override any of these requirements. But such consent must never be a necessary condition for access to a service like use of a search engine or making a restaurant reservation. All users must be guaranteed the option to purchase identical goods or services at their "shadow" price, based on the costs to the organization of providing them.

These are tough conditions. A right like this would threaten the viability of many for-profit, private-sector personal-data systems in America today. Most of what I have termed "unilateral" personal decision systems would fail this test—for want of consent from the subjects of the data. So would sales and trades of personal data aimed at targeting voters by political campaigns. Sales or trades of reports from consumer credit reporting agencies, or insurance reporting organizations, would require explicit consent from the subject of the data. Such consent would have to be much more specific and limited in scope than what credit reporting agencies collect today. For example, it should not be valid beyond a relatively brief contractual period—say, six months, with the option to renew.

The new right would protect all personal account data, including those held by retailers, credit card companies, banks and other financial institutions, charities, dating services, subscription-based periodicals, and websites offering medical advice. It would protect all data contained in prescriptions and medical records, and data on consumption habits related to health and illness—against sale or trade for commercial purposes. None of these frequently sought forms of personal information could be purchased or traded for commercial use without permission from the subject. Even the sale or trade of guest lists from hotels, motels, and resorts, with their sometimes-revealing implications concerning guests' sexual orientations and interests would be blocked, except with the consent of the subjects. In addition to sale and trade of all these kinds of information, the new right would block organizations from exploiting the troves of information on their *past interactions* with individuals to guide further dealings with those same people. That's right: the models of consumers' personalities and sensitivities developed through analysis of search-engine use or other intimate interactions could not be sold or otherwise exploited without consent from the subject. The parties that collect such data might face no obstacle to *recording* the data. But they could use such insights only with the consent of the subjects. And the granting of such consent could never be a condition for access to the organization's services.

What the new right certainly would *not* restrict are the vital exchanges of personal information involved in discussions of public affairs, civic values, and electoral politics—the broad realm of public discourse. Also unrestricted would be cash-for-trash journalism and mass dissemination of scandalous facts with an eye to influencing public events—so long as no sale or trade for value takes place. Nor would this right limit collection or dissemination of personal information for the writing of history or biography. The line of demarcation would always be the point at which any personal datum is sold to enhance the profit-seeking processes of any business, including polling companies and consultants offering to shape public processes for a fee.

Objections

Needless to say, any serious discussion of a right like this will trigger tectonic upheavals in relations between ordinary people and organizations now devoted to recording their lives. The subjects of such monitoring—all of us—would suddenly have something to offer to or withhold from the watchers: notably, consent to being reported on as well as a newfound ability to set terms for permitting any such use. Without such consent, users of such information would face either civil suits from aggrieved subjects or regulatory action by competent government agencies. The response from current users facing such restrictions on what the information industries have thus far treated as a low-cost raw material will, of course, be ferocious.

First Amendment Issues

We have already encountered objections that First Amendment enthusiasts are likely to declare against any curtailment of the right to disseminate true personal information under nearly any circumstances. Sometimes these thinkers propose letting "The Market"— that protean phantom of political economy—decide whose personal data should change hands, and on what terms. But before we begin to speak about markets, we need to know what is to be "marketed." If personal information is indeed the key source of value and advantage in the new world being born, who owns it in the first place? Nobody? Everybody? Or the persons whose lives it depicts?

Defense of "Innovation"

A second array of objections to property rights over personal information portrays such ideas as so many spokes in the wheels of Innovation. "Innovation" here figures as first cousin to Technology—

a beneficent if vaguely defined force in human affairs that appears highly vulnerable to interference by reservations on matters of principle. In this view, the United States is blessed in its freedom from regulations that might have blocked the brilliant successes of the GAFAM companies and their imitators over recent decades—thus permitting this country to foster some of the signal corporate success stories of the so-called Information Age. Enabling ordinary consumers to stand in the way of these developments, the advocates hold, threatens to stem the tides that are waiting to raise all economic boats. "If we constrict their fuel—data—we may hurt not only the quality, cost and speed of their services," writes one spokeswoman for this view, entrepreneur and investor Heidi Messer, "but also the drivers of growth for the world's economy."[7]

It takes a bold spirit to choose the role of cheerleader for innovation *in general*. Not all innovative departures lead to happy outcomes. DDT, thalidomide, and (more recently) Boeing's 737 MAX all won corporate enthusiasm at first, only to resolve into tragedy. Then there was that early twenty-first century array of innovations in the marketing of subprime and other speculative mortgage contracts hailed as brilliant at the time—yet whose poisonous repercussions are still being felt more than ten years later. And let's not forget the advanced thinking of the interwar Dutch governments in centralizing all their citizens' identity records in one site—complete with religious affiliation! Some innovations succeed too well.

Such reversals of public judgment in light of historical experience happen regularly in sweeping, technology-based programs of social engineering. One thinks of "urban renewal," or plans to dam the Nile and other large rivers—projects at first applauded by the foremost authorities, then later judged to have brought unacceptable environmental damage and social upheaval. With revelations like that of the once-concealed role of Facebook in attempts to sway voters in the 2016 presidential election, we see critical attention now directed at the entire American internet complex—accompanied by serious efforts to rein in these industries.

These historical comparisons may not move everyone. Enthusiasm for change—let me not say "Innovation"—in information technology has inspired many simply to leave the downsides of particularly bold "innovations" for others to worry about. But respect for minority concerns is essential to taking privacy seriously—in this case, the supposed minority of the public bracketed by Messer as "privacy evangelists." They, and anyone else who so chooses, deserve a right to "just say no" to participating in today's sophisticated personal-decision systems—that is, to reject the demands that these systems make on privacy, along with the supposed benefits yielded by them. Why should internet users who object to the tracking of their messaging and website searches not be guaranteed tracking-free options—if they are willing to renounce the dubious benefits of having their communications monitored and recorded for interested but unknown future parties?

Similarly, a grassroots privacy campaign should seek special protection for those who wish to drop out of the credit reporting system, or simply to take the equivalent of the Fifth Amendment on their credit histories. Some prospective creditors might be willing to consider such "no comment" reports, in conjunction with different forms of evidence of the applicant's ability to pay, as an alternative to permanent stigmatization of those with conventionally bad credit histories. By the same token, providers of services that promote themselves as "free," but that collect personal information on the user as compensation, should be required to provide identical services, also at their real cost, without appropriation of users' personal data and/or to state that, in exchange for "free" services, they collect the users' personal data. A right to "just say no" to being tracked by the institutions one deals with, like the right to resign from burdensome personal-decision systems, would ultimately help keep valuable alternatives alive.

Bad Associations

A final objection—this one shared by many privacy advocates—is more diffuse. The personal-data-as-personal-property idea seems to carry bad associations, in the form of an unwelcome libertarian ring. It is as though I were proposing simply to "let the market decide" what will happen to people's personal data. Too often, when one hears such appeals, the measures being promoted amount simply to leaving the resolution of complex public dilemmas to the tender mercies of the parties with the most resources.

But that is not my view. I don't believe in *The* Market—that mythic entity so often invoked without further unpacking—so much as in a multiplicity of particular markets, each with its own political content. Markets are human creations, such that planners need strong imaginations to define (a) *what sorts* of markets they seek to create and (b) how to shape the choices they afford and the social values they promote. The idea of property rights over commercialization of one's own data does open the way for some form of market in permission to use personal information—though no one would be required to participate in such a market. But it remains for thoughtful planners to decide how the limits of such property will be defined, what terms will be acceptable for purchasing access to personal data, for use over what periods of time, and so on.

Some fear that property rights over personal data could enable monied interests to purchase such rights *in perpetuity,* thus cementing permanent inequalities. One imagines, for example, private equity firms buying up the commercial data-rights of newborns at birth, from their low-income parents. Some version of such debacles occurred widely following the collapse of the former Soviet Empire: speculators snapped up shares in what were then failing enterprises at bargain-basement prices, from workers who little understood their value. Some of those unwise transfers seem to have paved the way for the current one-sided role of oligarchs in the ex-Soviet economies.

But well-informed planning for this new form of property can surely forestall such wrong turns in advance. Paul Schwartz, one of the most acute commentators on property theories of personal information, points out that the concept of property need hardly involve "despotic dominion over a thing," to use Blackstone's term.[8] Today we live in a world where information is shared, reproduced, withheld, analyzed, transformed, leased, and repurposed in countless ways. The problem we face is one of political wisdom, of knowing what legal and policy forms will best support privacy and other crucial public values without blocking uses that the subjects of the data might desire.

For the values animating this book, the possibility of selling commercial rights over one's own data definitively, in a once-and-for-all transaction, represents a step in exactly the wrong direction. It is much better to limit the period for which any citizen or consumer can assign such rights—say, for the purposes of this discussion, to six months, renewable by mutual agreement between the subject of the data and the user. Great care is necessary in specifying *which categories* of personal information and *what forms* of use would be permitted under any contractual agreement. But where consumers are willing to permit *any* commercialization of their data, the more precise and delimited the terms of such contracts, the better. And the new markets created (one hopes) by such conscientious planning need appropriate *institutions* through which to put the newly formed interests to work.

Institutions

The new institutions will be *personal data agencies*. These organizations will specialize in representing the wishes of consumers and, at their request, negotiating terms for release to data-using organizations. In playing this intermediary role, agencies will naturally seek to represent as many consumers as possible, while remaining attuned to demand from personal data-using organizations. In the course of this work, they will develop sensitivity to prevailing prices for personal data, and sophistication regarding terms to seek for their clients.

Agencies will learn to respond to consumers' wishes for discrimination in these respects. Some consumers, no doubt, will be willing to throw privacy to the winds, and simply seek the highest possible payout for indiscriminate release of their information. Others will welcome the new right and take advantage of it by doing nothing—hence switching off the onslaught of privacy-invading attentions and appeals that Americans today regard as normal. Still others will want to approve limited sharing of their information, while applying more or less precise conditions to its release. They may be ready to share their data only with nonprofit organizations, or women-owned businesses, or companies that follow right-to-work principles in hiring. Agencies capable of implementing such filters will benefit from their special appeal to specific categories of users—while presumably imposing higher commissions to meet their additional costs of fulfilling these special requirements.

These amount to a challenging set of constraints for the new data-rights industry, but hardly unprecedented ones. Like the long-established agencies that broker municipal bonds or soybean futures, the new industry will draw upon a certain cultural experience in bringing buyers and sellers together. Indeed, the new industry already has models in BMI (Broadcast Music Incorporated) and ASCAP (American Society of Composers, Authors and Publishers), the organizations that represent composers of music and other creators of content for public broadcast. Since the early twentieth century, BMI and ASCAP have monitored the public airwaves and live performances, billed the media that carried performances by their clients, and transmitted royalties to those clients. Like BMI and ASCAP, future data-rights agencies will need to act aggressively against those who use personal data commercially without permission—thus becoming enforcers of privacy obligations that ordinary consumers would find it difficult to monitor by themselves.

This new industry, then, will operate within a variety of desirable competitive tensions. Consumers—that is, prospective agency clients—will effectively seek representation by the "best" agencies— for example, those providing the highest royalty rates and most conscientious service. Agencies, for their part, will compete with one another for the most or the "best" clients—those who generate the highest volume of business, the highest prices for their data, and/or the most profitable demands for service. Presumably, the resulting industry will differentiate into various sectors—with some agencies catering to the needs of consumers with the fewest demands as to how their data are used, while others become "niche" players willing to implement discriminating demands from their consumer base— no doubt for a higher price. Regardless of the makeup of their clienteles, all agencies would find it natural to collect payments from data users on behalf of their clients, then issue royalty checks after each six-month cycle, deducting a percentage of the proceeds as their commissions.

Note two more things about the likely social profile of the new industry. First, the tasks it will face will be impossible without sophisticated and innovative uses of computing. The matching in real time of complex requirements of consumers for the handling of their data and the interests of the parties seeking to use the data, entirely feasible today, would have been miraculous two generations ago. Thus, computing could work to *enhance* control over one's own data (which would then truly be one's own), rather than eroding it. Second, creation of this new right will trigger the rise of new, productive forms of economic activity—including new businesses (above all, the agencies), new employment, new skills, and new sources of support for privacy values. Contra notions that regulatory intervention must be a drag on economic growth, the steps proposed here should help create a more vibrant economy—much as environmental legislation requiring solar panel production and careful recycling can contribute to prosperity.

How much of an economic difference would this new right make in the lives of American consumers? The question is eminently reasonable, yet any response requires educated guessing.

One published figure puts the number of unsolicited appeals received through the mail by the average American household at 848.[9] Suppose that this notional average family decided to impose a royalty of twenty cents on the release of the personal information leading to each of these 848 pieces. Suppose further that this added cost reduced the number of pieces received by one-third—a blessing in itself, most Americans would probably judge—to about 568 pieces, for a total revenue per family of 568 × $.20, or about $114.00. If the data-rights agency representing the family charged an annual service fee of 10 percent, that would still leave this strictly hypothetical family with a bit more than $100 income at the end of the year. To this, the family might seek to add revenues from permissions leading to email spam or unsolicited phone calls, which are even more speculative. Figures like these would hardly bankroll a lifestyle change for an average family, on these assumptions. But they might suffice for an extra holiday gift or charitable contribution at the end of the year. High-income families could probably afford to impose much higher fees for permission to use their information. But one suspects that many such families would be more likely to choose the privacy option of doing nothing, thus permitting no commercial use of their data.

Some privacy advocates nevertheless greet this scenario with skepticism. They may simply distrust reliance on market processes to help implement what they consider a basic human right. They might concede that data-rights agencies like those envisaged above could perform a useful function by helping spot and eliminate uses of the personal information of clients who have not granted consent for such use. But the critics could well argue that the agencies as described above could become a force encouraging Americans to release *more* of their personal information—which would, after all, mean more business for the agencies.

For some strong-minded privacy advocates, the underlying point is one of basic principle. Privacy should be a *right,* in this vision—not something that people are encouraged (or enabled) to sell. Any right that is up for exchange in this way, it's argued, amounts to an invitation to exploit those with fewest resources. From this viewpoint, it should also be illegal to sell blood, bodily organs, or services as a surrogate mother.

These arguments activate long-established lines of ethical debate, which I am certainly unable to resolve here. Let me just say that all sorts of reforms affect different sectors of the population to different degrees, depending on the privilege of their positions. How strongly one ought to oppose any given form of exchange ought to depend in some measure on *how bad* it is for a given actor to engage in that exchange. Selling oneself into slavery strikes most of us as bad enough to warrant outright proscription, partly because we count what the seller loses as particularly destructive of core values.

But other kinds of exchange, though undesirable, do not strike me as *that* bad. Unlike selling oneself into slavery, for example, they do not leave the decision-maker unable to change his or her mind. Freedom of expression is certainly a core value, and that value is undermined, to a degree, if people accept payment or other inducements to express opinions or sympathies they do not truly feel. But I would not favor legal sanctions against payment for such insincere expressions of views on public issues. This is partly because state inquiries into the sincerity of expressed opinions could readily become more repugnant than the insincere expressions themselves—and partly because I believe that the damage done by such expressions is considerably less, say, than that done by exchanges likely to cost people their lives or their health. My view of the evils of being tempted by payment for willingness to be bombarded by ads and all manner of other entreaties is much the same.

The overarching aim of this book is to encourage organized, grassroots support for privacy values—including certainly the value of extended control over information on oneself. It is hard to imagine launching such a movement while insisting to potential supporters that the sponsors of the movement-in-the-making know better than their prospective followers what uses of one's own personal data should be permitted. Such a stance would reek of paternalism. My hope is that American consumers, once new and meaningful choices are open to them, will approach such choices with skepticism and discrimination. At the very least, embracing the principle of *owning* the right to commercial exploitation of the details of one's own life should vastly raise the sophistication of Americans concerning what is at stake with such use *in general*.

Conclusions

If it were a taken-for-granted reality of everyday life that appropriation of personal information for commercial use without compensation amounted to theft—absent *explicit* consent from the subject—attitudes toward such activities would shift radically. Activities like geofencing would then become as comfortable an experience for the perpetrators as reaching into the subjects' pockets or purses and removing cash. This new cultural understanding would lead everyone to weigh the financial implications of capturing and exploiting personal data—whether for purposes of political campaigning, marketing and advertising, or credit allocation. Loss of control over data on oneself, instead of simply amounting to business as usual, would raise alarms like loss of a credit card or a hole in one's pocket.

Today, of course, we live in a world of calculated confusion about how information about ourselves is created, where it comes from, and how it affects our lives. Creation of the right advocated here will strongly countervail against these confusing and often paralyzing realities. It will help instill an understanding that access to and use of personal data of all sorts provides crucial advantages to organizations both public and private. It should, at a minimum, put an end to surreptitious privacy-invasion strategies like Geofencing. Everyone would understand that, where personal data were being compiled and held for commercial purposes, exploitation of such data without consent would be tantamount to theft.

At the very least, a new right like the one proposed here would leave no one worse off than today's status quo. Even consumers who made it their personal policy to sell all available data on themselves would be aware of what they were controlling, and aware of the uses being made of their information—which then would really be theirs. They would always retain the right to change their minds, even at short notice. The alternative course is more of the same where privacy is concerned—less control over data on oneself, more and more seamless exchange among holders and consumers of such data, and less public understanding of all these processes.

Isn't that a prospect worth militating against?

Notes

This chapter grew out of an essay published more than twenty years ago: James Rule and Lawrence Hunter, "Towards Property Rights in Personal Data," in *Visions of Privacy: Policy Choices for the Digital Age*, edited by Colin Bennett and Rebecca Grant (Toronto: University of Toronto Press, 1999). James Rule gratefully acknowledges the crucial and stimulating role of Larry Hunter in the original work.

1. Evan Hendricks, "Feds Pay $30 Million to Information Brokers," *Privacy Times* 26, no. 8 (Apr. 17, 2006).

2. Table 2 was produced by URAP students Angelica Vohland, Bani Bedi, Randy Cantz, and Han Cheng. The key sources for the data were the websites noted in the cells, all of which were last observed in January, 2021.

3. Federal Reserve Bank of New York, "Total Household Debt Increased in Q3 2020, Led by Surge in New Credit Extensions; Mortgage Originations, Including Refinances, Continue to Soar," Nov. 17, 2020, https://www.newyorkfed.org/newsevents/news/research/2020/20201117.

4. Drew FitzGerald and Sarah Krouse, "FCC Probe Finds Mobile Carriers Didn't Safeguard Customer Location Data," *Wall Street Journal*, Feb. 27, 2020.

5. Ibid.

6. Zack O'Malley Greenburg, "The Highest-Paid Dead Celebrities of 2016," *Forbes*, Oct. 12, 2016.

7. Heidi Messer, "Why We Should Stop Fetishizing Privacy," *New York Times*, May 23, 2019.

8. Paul M. Schwartz, "Property, Privacy, and Personal Data," *Harvard Law Review* 117, no. 7 (May 2004): 2056–2128.

9. Lauren Cahn, "This Is How You Can Stop Getting So Much Junk Mail—for Good," *Reader's Digest*, Jan. 13, 2020, https://www.rd.com/list/how-to-stop-getting-junk-mail/.

7   Conclusions

"Why take away the punch bowl," the question goes, "just when the party's getting started?" We hear this from disappointed investors, when the Federal Reserve raises interest rates as the economy gathers steam. Critics of the reforms proposed in this book will no doubt raise similar objections, for similar reasons. Why spoke the wheels of Technology and Innovation, they will lament, just when the economic triumphs and social amenities brought forth by the internet are flowing so abundantly? According to Heidi Messer, whom we have heard from earlier in this book, "The United States has been the architect of the new economy. But privacy evangelists have made villains of the very companies the world emulates. Rather than debate how to expand this economic opportunity, they call for fettering it."[1]

Like the exuberant figures quoted at the beginning of chapter 1, Messer blends a faith-based rendition of present-day realities with unreflective confidence in the future. Privacy is increasingly recognized as an anachronistic value, the message goes—and thank God for that! A new generation have shaken off superstitious concerns over the paths taken by "their" data. These upbeat moderns renounce such fruitless obsessions, in clear-eyed exchange for a cornucopia of material blessings and practical convenience. All this in the most dynamic sector of the world economy, the bountiful source of high-paying employment and—yes!—vast new fortunes. Why, then, do privacy evangelists insist on quarreling with success?

Partly because our definitions of *success* in this networked world are not, thus far, quite so unqualified as hers. And partly because her vision of what comes next strikes us as, let's say, fulsome.

Alternative Visions

This is not the place for predictions of specific events—the patenting of inventions, the passage of laws, the rise and fall of companies, or the like. Such prophecies often backfire, much to social scientists' chagrin. Christopher Evans's 1979 work *The Micro Millennium* was, and is, quite a good book. But it suffers from a cloudy crystal ball: "A car which refuses to start when its driver has ingested too much alcohol... could well be the only type on the road in the late 1980s. By then most cars will be theft-proof though... the rewards for car theft will be falling off markedly."[2] Statements like this help explain why many social scientists prefer to predict only things that have already happened.

But without predicting anything so specific as theft-proofing of cars, we can often spot broad trends, future tensions, emerging institutions, and relationships that seem likely to continue to shape human affairs—or to slip from the historical stage. Thus, refinement in what social scientists bracket as *mass surveillance* and *social control* appears likely to continue shaping the world's "advanced" societies for the foreseeable future.

Imagine a radically ambitious program of this sort in the United States, aimed at monitoring the movements of the populace. It would enable the authorities to track and record the whereabouts of every American resident in real time, all the time. Earlier chapters of this book have detailed technologies that could provide the groundwork for such new surveillance forms. These include Stingray, already in wide use among American law enforcement agencies to track the movements of cell phone users, and geofencing, which identifies the users of cell phones simply by scanning crowds in public places. Judging by the increasing use of these technologies and the ingenuity already driving similar innovations in the United States and abroad, comprehensive tracking of all Americans' whereabouts and movements in real time could soon become a practical possibility.

Results of this monitoring could feed into a national center managed, under the highest security, by a team of experts who would provide liaison with local, county, state, and other levels of law enforcement. To reassure the public, one can imagine confiding this responsibility to seasoned senior officials—former high-ranking FBI agents or FISA Court judges, for example, or experienced managers from the National Security Agency.

Cell phone use, in this new regime, would be necessary to conclude every major financial transaction—every retail purchase, every road or bridge toll passage, every rental payment, and every mass transit ride, every medical care visit, and all the other transactions that make up everyday life. Each of these encounters would leave its record in the unfolding timeline of everyone's account. That comprehensive record would offer all sorts of broad social benefits. Whenever a crime was reported, for example, the authorities on the ground would review the historical record to identify those present at the scene—making crime an unattractive option for any rational actor, and apprehension of the guilty parties all but automatic. Missing persons would virtually become a thing of the past. Persons under suspicion of being past perpetrators—or, for that matter, likely future perpetrators—of crimes could be followed unobtrusively until the authorities had grounds to act more forcefully. Those found by the authorities to be outside their home regions could be called upon to explain their presence. Prosecutions of those charged with crimes would virtually prepare themselves, based on recent surveillance of the people and places concerned.

True, there would be resistance—as the recent grassroots opposition to COVID-19 prevention measures demonstrates. Yet if this initiative could continue long enough to show positive results, a constituency of support could well develop. That of course was the counsel of Thomas Hobbes, the austere English philosopher, who saw deep wisdom in citizens' offering their law-abiding compliance in exchange for protection from the nearest regime capable of providing it. If people came to see government tracking of everyone's movements as a source of relief from anxieties about crime and other uncertainties, invasive new forms of surveillance could win a significant following. Some Americans would resist some superficial inconveniences—for example, the need to have their phones with them constantly, and to use them for nearly every transaction. But for most law-abiding citizens, the rewards of a system like this could constitute a major social dividend.

Success could open the way for many similar programs of social improvement through intensified state-sponsored monitoring. A project aimed at improving Americans' health, for example, could apply helpful "nudges" to the public's diet, and the amounts and quality of the physical exercise they engage in. This would involve scrutinizing every purchase at the grocery store, and every movement undertaken in the course of a day, monitoring blood pressure and pulse rates, and checking the frequencies of social contacts and other health-relevant data. Once again, use of a cell phone or implanted computer chip for the approval of credit or debit purchases could facilitate persuasive encouragement to make good choices in these respects. By reducing the costs of expensive diseases resulting from unhealthy lifestyles, a program like this might quickly pay for itself.

Similarly, hardly anyone could deny that America would be a better place if we all voted at higher rates in local, state, and national elections—or if we informed ourselves more thoroughly about matters before the public and communicated more thoughtfully with neighbors and fellow citizens on such matters. If we indeed share these goals, perhaps we should not hesitate to apply the full capacities of computerized surveillance to attaining them. In addition to credit scores, Americans could also accumulate Civic Mindedness Scores, maintained by a federal agency devoted to such reckonings. People could receive credit for donating blood, performing community service, and offering constructive inputs to public debate on issues of the day. Notably antisocial behavior like littering could result in points off. Running calculations of citizens' current scores could be made available to responsible government and private organizations at their request.[3]

One dangerous state of affairs that would tempt nearly any regime to intensify monitoring of its population is the spread of infectious disease—as in the coronavirus pandemic that began in Wuhan, China, in 2019. There the government reacted as one might expect of a technologically sophisticated state little troubled with legal restraint on political power. And there, too, cell phones played a key role in efforts to combat the disease: during the pandemic, citizens were classified by three color codes—green, yellow, and red—based on the history of their movements. These codes were available via each person's cell phone, which had to be used for crucial transactions. When seeking access to high-speed rail service or bus connections or when entering other public spaces likely to be crowded, everyone expected to have their codes checked. Those coded red or yellow were assumed to have had close contact with a confirmed coronavirus carrier; the authorities would block holders of these phones from further travel and possibly assign them enforced quarantine. Those lucky enough to show a green classification faced few restrictions. An additional feature of the system enabled the authorities to drill down into the details of the person's traveling history, should suspicions remain.[4]

Going Too Far?

In fact, surveillance *infrastructure* now extant or nearing development in the United States could well support more effective strategies than China's for control of grassroots irritations. Most of China's government mechanisms for tracking individual behavior are not centralized nationally but held in regional centers—whereas Americans are covered by credit records, data brokerages, FBI identification systems, and many other systems that are organized centrally.

American readers might well challenge the comparison of these latter organizations to databases usable for repression. But consumer credit reporting has become a gatekeeper for many connections beyond retailers that extend credit to their clienteles.[5]

The extraordinary evolution of American credit reporting to cover the lives of nearly all this country's adult population—and not just their credit accounts—has made it an irresistible target for co-optation by national security interests. Since the 9/11 events, federal authorities have commandeered this system as a mechanism for identifying and pressuring individual citizens as supporters of terrorism or other illegal activities. According to a report from a lawyers' group:

> The Office of Foreign Assets Control [of the Treasury Department] keeps a list of over 6,000 businesses and individuals that they deem suspicious of terrorist actions. Before 9/11 this list was only in the hundreds, but has since grown as the criteria widened. Companies and businesses are prohibited from doing business with those on the list unless authorized by OFAC. Anyone on that list can be blacklisted from conducting day-to-day business. . . .
>
> This happened to Amit Patel, who was unable to find a job for two years; which he later found was likely due to the false terrorist tag on background checks. A man that had a 700 credit score, he was even

denied an apartment. When he confronted his landlord about this, he was forwarded his credit report which is when he found out that he was labeled as a terrorist.

As it turned out, the Amit Patel bracketed as a terrorist was someone with the same first and last name, but not the same person as the one who suffered the relentless persecution described here. The authors of this account drily comment: "The Treasury Department admits that the OFAC list generates many false positives [cases where people are wrongly identified as terrorists]. This is tragic, because many Americans have been denied jobs, financing, and even health insurance because of an OFAC alert."[6]

These stunning powers to shut down the very livelihoods of American citizens and residents derive from Section 302 of the Anti-Terrorism and Effective Death Penalty Act of 1996. This statute allows the secretary of state to designate foreign terrorist organizations (FTOs) and makes it a crime for U.S. persons to provide an FTO with material support.[7] It requires U.S. financial institutions to block funds of designated organizations and persons.

Legislation like this could be taken from a guidebook for the conduct of authoritarian government.

Some readers might see Amit Patel's plight simply as the result of a good-faith mistake by Treasury officials. But no one would draw that conclusion from the case of those targeted in the following account. This report, based on a *New York Times* article from February 2020, details an FBI campaign against three truck drivers of Pakistani origin.[8] Early in the new millennium, FBI agents approached Muhammad Tanvir with a request that he provide information on fellow Muslims in his community. A lawful resident of Queens, New York, Tanvir declined, on grounds that his faith forbade such conduct. The agents' pressure continued over months and finally, in 2010, they raised the ante, placing him on the national no-fly list, which barred him from boarding flights leaving from or arriving at American airports. Since his work often required him to drive to distant points and return by air, this sanction proved disabling. The agents assured him that he would be removed from the list if he cooperated with their inquiries.

Tanvir quit his job. Three times, he bought plane tickets to visit his mother in Pakistan, who was in failing health. He was blocked from flying on each of those occasions. The agents continued their pressure on him to inform on his co-religionists. Here the authorities evidently knew exactly what they were doing in applying these illegitimate but extremely forceful measures against Tanvir and his associates. Yet those applying the pressure must have calculated that nobody would draw attention to this rough justice for what it was.

Tanvir was fortunate enough to find good legal representation from Ramzi Kassem, who took this case, joined with those of two other Muslim men subjected to similar FBI pressures, to the Supreme Court. "Our clients were placed on the no-fly list not because they posed any threat to aviation security," Professor Kassem said, "but as a way to coerce them to become informers on their own communities."

Results of ensuing court actions were revealing. In 2018, the Second Circuit of the U.S. District Court for the Southern District of New York blocked a petition from the three to sue the FBI agents involved in their ordeal. But the following year, the Supreme Court agreed to hear their case, and in December 2020 the Court upheld their right to seek damages from the agents by a margin of eight to zero. The justices based their decision in part on the Religious Freedom Restoration Act of 1993.[9] The plaintiffs' success here stands in contrast to the experience of Onree Norris, the seventy-nine-year-old victim of the 2018 no-knock raid in rural Georgia recounted in the preface. His attorney's attempt to sue the invading officers was denied on the grounds of "qualified immunity" of public officials in the exercise of their responsibilities. The fact that Tanvir and his associates were thought to have religious motives for their resistance to the FBI seems to have provided a crucial advantage. Here there was no mistake. The federal authorities evidently knew exactly what they were doing in applying this illegitimate but extremely forceful pressure on Tanvir and his associates.

The story of Muhammad Tanvir and his associates holds multiple lessons.

The first is that some key legal protections seem to have withstood the tensions of the times. The fact that the three men gained access to strong legal representation, and that the politics of the then current administration in Washington had not totally gained ascendance over the courts, apparently saved them from what would have been intolerable pressure from the FBI. In the forum of public opinion, the fact that the three were Muslims might have disqualified them from consideration.

Yet the story also demonstrates the far-reaching bite of the *mechanisms* available to the authorities for exerting pressure on ordinary citizens. Here the news regarding surveillance in America is not so heartening. The threats reportedly directed against Tanvir and his associates were apparently available strictly at the discretion of the FBI. Like the devastating sanctions applied to Amit Patel via his credit record, their operations were designed to be opaque to those targeted in them. And these highly centralized mechanisms are organized to work systemically, so that once the figurative switch is flipped in Washington, the constraints are felt wherever in the United States one seeks to board a scheduled flight, or wherever one needs a credit report. Here, as in the case of no-knock warrants for armed home invasions, we have the civil equivalents of "weapons of mass destruction"—instruments of power too sweeping to be accorded to those entrusted to enforce the law.

Despite the availability of enforcement capacities like these, many American readers may doubt the latent potential for repressive uses of their fellow citizens' personal information. Deep-seated privacy sensibilities, and habits of resistance to overbearing government, they will hold, grant Americans long-entrenched immune reactions against authoritarian policies.

It would be reassuring to think so. But one suspects that what Americans can be persuaded to accept depends decisively on the official framing of what is at stake—rather than on some absolute level of intrusiveness. I was fascinated, early in the new millennium, to see reports on "backscatter" and similar devices for use in screening passengers in U.S. airports. These machines are capable of photographing people "naked" in street clothing as they pass through airport security checkpoints. The resolution of the resulting images is claimed to be precise enough to reveal whether a given male traveler is circumcised or to disclose the size of a woman's nipples.

Surely, I thought, the forces of surveillance were finally overplaying their hand. However highly Americans might rate the demands of the so-called war on terror, this proposal marked a bridge too far. Having photographs of one's unclothed self, recorded and scrutinized by agents of one's own government—surely this prospect would cause most Americans to balk. And yet, American travelers acquiesced by the millions to the backscatter scans during airport screenings from 2010 to 2013, until the technology was replaced with machines that produced only slightly less revealing images.[10]

Public attitudes on surveillance appear to be at least as malleable as those on same-sex marriage, racial equality, or religious observance—all of which have experienced dramatic and often abrupt change over recent decades. What information people consider reasonable to yield to which outside parties evolves constantly. Demands for personal information associated with income taxation and Social Security, in the first half of the twentieth century, triggered significant public ambivalence and resistance. Pro-labor groups, for example, feared that Social Security numbers would be used to track union sympathizers as they moved from job to job—and hence to discriminate against potential organizers.[11]

These expressions of concern brought government assurances that the Social Security program would remain an independent data-keeping operation whose personal data would not be shared with other organizations.[12] But today, sharing of Social Security data is widespread for a variety of extraneous purposes—including, for example, pursuit of parents who abscond on child-support obligations.[13] The point here is that bureaucratic instruments for tracking and enforcement, once in place, readily become virtually irresistible to other organizations or agencies with quite different enforcement interests. If Social Security surveillance, for example, appears to offer the only way of enforcing legally valid court orders to obtain child-support payments, refusal to cooperate may become politically impossible for the Social Security Administration.

How Do "Societies" Decide?

How do major realignments of public conviction on such key matters of principle come about? What has to occur, for public opinion on matters like same-sex marriage, abortion, or racial justice to take place? Are these shifts best understood as the direct aggregate results of many individuals' strictly independent decisions? Or are more complex processes at work, by which public opinion and public action influence one another?

No one has written more incisively on this subject than the late Cornell economist Alfred E. Kahn—above all, in his classic article "The Tyranny of Small Decisions" (1966).[14] Kahn focuses on the example of train service to and from his home in Ithaca, New York—where winter weather often rules out access by air and road. The city had once had good passenger rail service, Kahn points out. But rail travel to Ithaca gradually lost market share to car, bus, and air travel, to the point where the railroads discontinued passenger service altogether. This loss stemmed from countless theoretically rational, independent decisions by individual travelers to choose alternatives to railroad service. The cumulative result, Kahn argues, was one that perhaps no one would have chosen, had the choice-making been structured properly.

"The fact is," Kahn wrote, "the railroad provided the one reliable means of getting into and out of Ithaca in all kinds of weather, and this insufficiently exerted option . . . was something I for one would have been willing to pay something to have kept alive."[15] The point, once Kahn explains it, is devastatingly simple. What people were "choosing," when they made the decisions that led to loss of passenger rail service to Ithaca, was not what they might ultimately have cared about the most—that is, their option of a range of possible choices sufficient to ensure guaranteed transport in and out of Ithaca year-round.

Kahn is intimating that the language of economists fails us when it encourages us to regard established social practice as evidence, ipso facto, of what the public "really" wants—in cases like networks' choices of TV programming, for example, and other cases where people seem to embrace unsatisfactory social outcomes. Think of a suburban homeowner who complains about congestion on the main highways between her home and her work. We should note that what this economic person may have "chosen" in the first place could have been an image of all the mythic images of suburban life—a spacious dwelling, quiet streets, plenty of parking, and good travel conditions. It is entirely possible that such an image influenced enough people to move to what had been the country to render itself out-of-date by the time the homeowner arrived in her supposed utopia.

Kahn goes on to reframe other momentous public "choices" that have been made through such inappropriate nondecisions—for example, those involving America's massive dependence on cars.

He quotes philosopher Morris Cohen: "Suppose.... Some being from outer space had made us this proposition: 'I know how to make a means of transportation that could in effect put 200 horses at the disposal of each of you. It would permit you to travel about, alone or in small groups, at 60 to 80 miles an hour. I offer you this knowledge: the price is 40,000 lives per year.' Would we have accepted?"

Would we?

A Defeat for Privacy

Kahn's analysis has much to teach us about real-world struggles over privacy. Consider an extended public deliberation on surveillance—in Lockport, New York, a city of about twenty thousand in Niagara County—that did not yield a privacy-friendly outcome, and where the defeat of privacy values paralleled that of winter train service in Kahn's Ithaca. The case comes from a detailed and thoughtful *New York Times* article by Davey Alba:

> Jim Schultz tried everything he could think of to stop facial recognition technology from entering the public schools in Lockport....
>
> But a few weeks ago, he lost. The Lockport City School District turned on the technology to monitor who's on the property at its eight schools, becoming the first known public school district in New York to adopt facial recognition, and one of the first in the nation....
>
> Robert LiPuma, the Lockport City School District's director of technology, said he believed that if the technology had been in place at Marjory Stoneman Douglas High school in Parkland, Fla., the deadly 2018 attack there may never have happened.
>
> "You had an expelled student that would have been put into the system, because they were not supposed to be on school grounds," Mr. LiPuma said. "They snuck in through an open door. The minute they snuck in, the system would have identified that person."...

When the system is on, Mr. LiPuma said, the software looks at the faces captured by the hundreds of cameras and calculates whether those faces match a "person of interest" list made by school administrators.

That list includes sex offenders in the area, people prohibited from seeing students by restraining orders, former employees who are barred from visiting the school and others deemed "credible threats" by law enforcement. . . .

The technology will also scan for guns. The chief of the Lockport Police Department, Steven Abbott, said that if a human monitor confirmed a gun that Aegis had detected, an alert would automatically go to both administrators and the Police Department. . . .

Mr. LiPuma, the director of technology, . . . said he, as well as some other school officials, would like to add suspended students to the watch list in the future . . . despite the State Education Department's recent directive that Lockport make it clear that in its policy that it is "never" to use the system "to create or maintain student data." Most school shootings in the last decade, Mr. LiPuma said, were carried out by students. . . .

"The frustration for me as a technology person is we have the potential" to prevent a school shooting, he said. . . .

Opponents of the new technology now pin their hopes on state lawmakers. In April, Assemblywoman Monica Wallace . . . introduced a bill that would force Lockport to halt the use of facial recognition for a year. . . .

"We all want to keep our children safe in school," she said. "But there are more effective, proven ways to do so that are less costly." . . .

Mr. Schultz said he would keep making his case.[16]

Note the position staked out by Robert LiPuma, this drama's self-described voice of "technology": tragedies like the massacre of students in Parkland, Florida, are preventable by technologies like the one proposed for Lockport. Shouldn't any caring parent or taxpayer be willing to go to any feasible length to forestall another event like that?

Facing a challenge like this, privacy advocates often find it hard to avoid going on the defensive. Before even entering into the substance of her objections, Assemblywoman Wallace felt called upon to affirm her own bona fides: "We all want to keep our children safe in school," she insisted.

This was not exactly a counterpunch from a position of strength.

Wallace apparently felt that she could not afford to allow the implication that she was indifferent to the fate of slain schoolchildren to go unchallenged. She then went on with arguments often heard from privacy advocates in her position: other measures, other expenditures of public funds could be more *cost-effective* in preventing gun violence in schools. Reasonable observations, all of them. But once it's accepted that the ability to *forestall* violent attacks on schoolchildren at school represents the ultimate standard of success, privacy forces have surrendered the high ground.

In fact, armed attacks on children at school are rare among all sources of death for children. A report from National Public Radio quoted University of California, Davis, researcher Garen Wintemute as stating, "Schools are just about the safest places in the world for kids to be.... Although each one of them is horrific and rivets the entire nation for a period of time, mass shootings at school are really very uncommon, and they are not increasing in frequency."[17] In light of the relative rarity of school shootings, a stronger response to the Lockport project might have been to compare the very hypothetical danger of death through firearms at school with the shared costs of taking one more incremental step toward a destination that hardly anyone really wants to reach.

I suspect that every privacy advocate can recount his or her own version of the Lockport debate. Virtually no one, in these exchanges, comes out *against* privacy. Instead, discussion devolves into a contrast between what is presented as a concrete *emergency*, threatening real people with vivid and dire consequences, and abstract and diffuse privacy concerns. In Lockport, the first of these alternatives was portrayed in terms of the combined forces of well-armed, ruthless school assailants, ready to prey on schoolchildren at any moment. The other value at stake—privacy, however understood—must have seemed far less pressing than the images of suffering, innocent schoolkids—even if there seemed to be no reason to suspect such an attack in that town in particular.

Things would be different if privacy advocates could draw on vivid, widely shared images of a world utterly without information privacy—and contrasting images of a world guided by reforms like those entertained here. If privacy—that diffuse set of values and concerns—acquired the high profile of, say, equal opportunity for women and minorities; or support for efforts against climate change; or the need for college opportunities for recently released prisoners, *then* arguments like the one waged by Jim Schultz in Lockport would become harder to ignore. I am thinking of images with the impact of polar bears swimming for their lives as Arctic ice floes melt (as bases for climate change appeals); or of Cleveland's Cayuhoga River ablaze in 1952 (as a spur to ecological activism); or of the deserted city of Chernobyl (for opponents of nuclear power); or of Seattle residents in the summer of 2021 frying eggs in skillets, without stoves—simply by leaving their skillets in that city's suddenly blazing sunshine. Without such transformation, that basket of distinct but related concerns flying under the flag of privacy is apt to play the role of assured train service to Ithaca, New York, in Alfred Kahn's example—a distant and abstract concern that everyone prefers to think about later.

If sperm whales, endangered wetlands, bluegrass music, and disappearing languages can all draw significant numbers of dedicated movement supporters, shouldn't privacy values attract at least as much public attention and energy? Appendix 2 shows an array of mobilizations of popular participation on behalf of privacy—"privacy affirmations"—on matters ranging from national identity cards to electronic tracking of the movements of elementary schoolchildren while on campus. If there is anything troubling about the events described in that appendix, perhaps it's the uncertainty as to how many lasting changes remained, after the privacy-affirming mobilizations in question. But the California voters' initiative on privacy (discussed below, and case 24 in appendix 2) seems likely to reshape privacy politics for some time.

A (Qualified) Privacy Victory

Let us agree that the big players in America's information industry enjoyed a long honeymoon in American public opinion. For Heidi Messer and her followers, those companies represent cornucopias of growth in employment, profits, and endless streams of beguiling new conveniences and creature comforts. For many other Americans, including some of the most sophisticated commentators, they embody that happy ideal of doing well by doing good. GAFAM (Europeans' not-always-appreciative nickname for Google, Amazon, Facebook, Apple, and Microsoft) seemed to combine the Enlightenment virtue of applying scientific analysis to real-world problems with the ability to more than amply reward their investors and employees. The perfect search engine, Google's Sergey Brin famously averred, would be "like the mind of God"—disclosing an extraordinarily far-reaching ambition, even for a founder of one of the world's most successful companies. Google's aspirational slogan, "Don't be evil," seemed to many no more than consistent with the company's early trajectory.

In an acute bit of reportage on this period, Nicholas Confessore wrote in the *New York Times:* "To many people in Washington, they were the good guys. Through the Obama years, the tech industry enjoyed extraordinary cachet in Washington, not only among Republicans but also among Democrats. Partnering with Silicon Valley allowed Democrats to position themselves as pro-business and forward-thinking. The tech industry was both an American economic success story and a political ally to Democrats on issues like immigration. Google enjoyed particularly close ties to the Obama administration: Dozens of Google alumni would serve in the White House or elsewhere in the administration."[18]

But winds of change were blowing. In June 2013, Edward Snowden burst from near-total obscurity with enormously detailed revelations of highly invasive government spying on ordinary Americans. Much of this activity involved tapping telecommunications lines—phones, email, internet, and the like. And a handful of the country's information-industry giants, including virtually all the major telecommunications providers, proved to have willingly opened their information streams to the hungry attentions of the NSA. Many close observers of this scene—in Washington and in the industry—did not seem deeply surprised at the facts revealed by Snowden, though many were flabbergasted that one figure could succeed in exposing such a broad swath of officially secret activities.

For some, no doubt, the extent of penetration by internet companies into the lives of ordinary citizens and consumers represents mainly an aesthetic issue—evoking an instinctive revulsion against being trailed, monitored, and anticipated. To those whose response went no further than that, Silicon Valley had a ready riposte: if no *harm* can be shown to result from these intrusions into once-private realms, then surely no damage has occurred. But the issues at play go beyond squeamishness.

Of course, the fortunes of the information-industry giants benefit enormously if consumers and their representatives must prove "harm" in order for courts to take any interest. Recall the revelations by Brian X. Chen of the *New York Times,* who found that more than one hundred companies knew about his buying habits and personal life via his Facebook page—mostly without his awareness. He may well have been disadvantaged in transactions resulting from the tracking that he described. He might, for example, have been directed to products and services that had high profit margins, rather than to ones that represented his preferred brand or style. But any consumer who set out to prove such treatment would presumably have to demonstrate the loss by comparing what they were directed to and what they would otherwise have chosen. Such personal loss, though certainly possible, would prove painstaking and onerous to demonstrate.

The San Francisco Bay Area was and is, of course, home to the largest single concentration of propaganda machinery on behalf of the information industry. But by the 2010s, a critical mass of independent thinkers had gravitated to the region as well, and not all the ideas emerging from that culture were reassuring to the industry. One of these articulate voices was that of Ashkan Soltani, a computer scientist and former chief technologist at the Federal Trade Commission who had an in-depth acquaintance with the relations between the industry and its customers, along with a gift for unsentimental phrasemaking on the subject. "It's like selling coffee and making it your job to decide if the coffee has lead in it," he remarked to me. "When it comes to privacy, we have no baseline law that says you can't put lead in it."[19]

By the middle of the decade, such sentiments were being heard more and more widely among commentators on the internet. But the next game-changing step came not from an internet intellectual or known civil liberties advocate, but from a wealthy real estate developer who seems to have stumbled across privacy concerns more or less by chance. Alastair Mactaggart, born in the Bay Area of California in 1966, had continued his family's real estate business. The Bay Area was an auspicious place for such activity in the late twentieth and early twenty-first centuries. With lively intellectual curiosity and ample means to pursue it, he began to read into the growing number of accounts of what the institutions he dealt with over the internet could know about him and his life.

In 2017, Mactaggart and investment strategist Dick Arney began to consider legislation that would establish rights they saw as seriously lacking in the United States. They soon set their sights on a typically Californian strategy: a statewide ballot initiative. By going directly to the state's voters, they would avoid the pitfalls of horse-trading in Sacramento, where many a populist initiative has foundered on the shoals of lobbying efforts and backroom politics. And ballot initiatives in California, if successful, are constitutionally very difficult to overturn. Mactaggart and Arney hired a small staff and set about collecting the roughly three hundred thousand signatures needed to get their measure on the ballot.

A key figure among this small working group was Mary Stone Ross, who coauthored the ballot measure. Like Mactaggart and Arney, Ross was an unpredictable figure as a privacy advocate, with previous appointments as a CIA officer and as counsel to the House Intelligence Committee. One of her responsibilities in the latter position had been oversight of NSA operations later disclosed by Edward Snowden. Regarding the latter, she later wrote that she found uses of personal data in the private sector "many times more intrusive [than government surveillance], and yet—in contrast to the NSA program—there was very little oversight."[20]

The ballot measure attracted wide support. It focused mainly on large businesses, which would be required to disclose to any consumer who inquired most of the information they had collected on him or her. The businesses would also be required to disclose "inferences" they may have drawn about their data subjects—say, regarding tastes in recreation, food, or values. Customers who didn't like what they discovered could opt out of such recordkeeping altogether—a choice, Mactaggart and Arney were convinced, most Californians would be eager to embrace. The effect of any such measure, of course, would be devastating to the industry in the form that it had taken since the early 2000s. After all, the ability to collect such otherwise "wasted" information and fashion it into bases for influencing their subjects' behavior is central to their model.

In 2017, Ashkan Soltani joined forces with Mactaggart and Arney. Soltani had broad experience in privacy politics, both as a civil servant in Washington, D.C., and in Illinois, one of the few states that had adopted legislation imposing significant restrictions on corporate use of personal information. The campaign to put their measure on the ballot was doing well, supporting their earlier suspicion that the California public was more than ready to see limits placed on the sorts of corporate sharing of personal data that had so disturbed Mactaggart. But Soltani fully expected Silicon Valley to put up a major fight against the measure, judging from the importance of the restrictions it promised and the resources available to the industry.

He was not disappointed. At first, the big information companies sought to launch alternative legislation to Mactaggart's ballot initiative in the state legislature in Sacramento. Then, when the insurgents filed petitions on behalf of their ballot measure bearing the signatures of nearly 630,000 California voters, and it began to look as though the impetus of the measure was overwhelming, the opposition suggested compromise. Mactaggart and his allies began discussions on a compromise bill that, he was convinced, would retain most of the provisions of the ballot measure. But as the days wore on, he saw these negotiations slipping into a quicksand of Sacramento politics. Some of what he considered essential elements—for example, the right of aggrieved consumers to sue companies for violating requirements of the law—appeared to be targeted for elimination by industry interests. He began to edit a compromise version of the bill. And he challenged the industry lobbyists now arrayed against his original position: they could agree to his proposed compromise in the legislature—or face the prospect of victory for his original ballot measure the following November. At this point, the opposition folded and the state legislature passed the California Consumer Privacy Act (CCPA) at virtually the last possible moment, with no dissenting votes. Governor Jerry Brown signed the measure into law on June 28, 2018. It went into full effect in the summer of 2020.

The CCPA indeed mandates far-reaching change in treatment of personal information. Its targeting of specific institutions and specific information processes within them for reform, however, is quirky at best. It applies, first of all, to businesses, and rather large ones—those with more than $25 million in sales or fifty thousand customers—though, almost as an afterthought, it includes any business "earning over half its annual income from selling customers' data."[21] At the same time, the law bypasses altogether the activities of government agencies, smaller companies, not-for-profit organizations, and many publishers—all of which have been targets of data-protection legislation elsewhere.

The CCPA establishes the right of all customers of the selected businesses to scrutinize the records held on them by the company. It guarantees these consumers a right to learn of any "characterizations" of them made by the company in its communications with other parties concerning the consumer. Presumably this refers to nicknames used by companies like InfoUSA, which sold data on ninety-two-year-old army veteran Richard Guthrie: "Elderly Opportunity Seekers," "Suffering Seniors," "Oldies but Goodies." Most importantly, CCPA also creates a right to opt out from sales of data on oneself and establishes that no discrimination must be applied to any consumer for exercising the new rights established by the law. The opt-out provision, as every privacy advocate will note, is far weaker than proposed opt-in language, which would make consent necessary for any disclosure. The law proscribes sale of personal data but does not directly address "sharing" of such data among organizations—to cite the winsome term often put forward by the industry for their activities. Perhaps needless to say, legal battles over such ambiguities erupted nearly as soon as the law went into effect.

Proposition 24

Supporters of the CCPA had always included a strangely heterogeneous cast of political types. Mactaggart himself, the prime mover, is a lifelong businessperson of independent views, but he had enlisted support from groups ranging from privacy and civil liberties activists to some figures from the information industry. Like many another California coalition, this one fell apart immediately after the CCPA victory in 2018. Mactaggart and a number of allies went back to the California electorate in 2020 with a new ballot measure, Proposition 24. Sold to the public as a consolidation of consumer protections in the earlier Act, Proposition 24 struck many as a disastrous retreat.

Thus, the Proposition eliminated the private right of action—that is, the right of consumers to sue keepers of their data for failure to adhere to provisions in the Act. Many long-term privacy advocates decried this step as capitulation to the industry. The American Civil Liberties Union, among others, objected to provisions that sweepingly exempted the consumer credit reporting industry from the new law. They also took great exception to the end of the "global opt-out" provision that would have permitted Californians to block the sale of their data with a single, definitive refusal. Others opposed provisions that, they held, would permit discrimination by companies against consumers who exercised the right guaranteed in the Act to opt out of release of their data. There were, and continue to be, major disputes and confusion over how the Act would be enforced. One of its most articulate opponents was Mary Stone Ross, who had coauthored the original ballot proposition with Mactaggart in 2018. Given that the Proposition—not the 2018 Act, but the 2020 ballot measure intended to revise it—came to some fifty-two pages, these conflicts seem destined to form the context for privacy politics in California for some time.[22]

Nevertheless, Proposition 24 passed by a comfortable margin in November 2020.

Despite the unfortunate revisions contained in Proposition 24, and despite many anomalies and flaws in the 2018 Act, I count the CCPA as a historic innovation—and one likely to shape the further development of personal information law in the United States well into the future, with or without further support from Mactaggart. The distinctive thing about the legislation is its implicit assumption about the underlying interests of data subjects *versus* data-keepers. Gone is the notion that institutional data-keepers and individual consumers form part of the same community of interest. Implicit in the language of the law from the beginning is the notion that many treasured institutional uses of customers' personal data were detrimental to basic interests of the consumer. It thus states out loud what some official voices have stated only in whispers. The law does not require the latter to demonstrate damage or harm through the sale of their data. Instead, it starts from the premise that it is normal to want to retain control over data on one's own transactions and business relationships—and that the sale and resale of such data are likely to undermine that interest. In effect, California law invokes privacy as a complex of values distinct and important enough in their own right to warrant upholding against a major sector of the consumer economy.

The CCPA appears to have triggered changes that the information industries and their many political allies will find it difficult to stop—hence the intensity of efforts by giants of the information industry to block the extension of CCPA-like measures on a national scale.[23] This book has consistently lamented the *asymmetry* of relations between ordinary citizens and consumers and the institutions that systematically track them. The passivity, indeed fatalism, that most Americans seem to experience concerning their records is both a bad thing in itself and a source of blockage of needed change. For its potential to provide consumers with tools to shape the fate of data about themselves, I regard the CCPA as a vast step ahead. It provides ordinary Californians an opportunity to turn away from the role of objects of inscrutable forces. The experience of being able to "just say no" to these processes, I believe, will change the mind-set of Californians, and nudge them and other Americans into more active roles in the managements of "their own" data.

The Reforms

The CCPA represents a shot across the bow of this country's surveillance institutions. But it addresses only a small subset of the broad concerns raised in this book. These reforms, of course, apply to large personal-decision systems of virtually all kinds. If they have any special usefulness, it lies in their mutually reinforcing, synergistic quality. Taken together, they would promote the following broad directions of change:

- All the proposed reforms aim at something more fundamental than simply *adding privacy* or *correcting abuses* within existing systems. It will not do, I hold, to start with the assumption that every personal-decision system that exists corresponds to an authentic *need*—in the sense of supporting widely shared common interests. Some personal-decision systems appear to serve no interests other than those of the entities that create and maintain them. Accordingly, I have proposed that many systems that do not enjoy consent from their subjects should be shuttered for lack of legal bases. Many other systems should be called into question simply because the organizations maintaining them are unable to provide a coherent account of why they are *needed* to complete any process or transaction sought by the subject.

- All the proposed reforms rely in the first place on people's own judgments of the acceptability of the personal-decision systems confronting them—hence, the right to withdraw from systems where participation is not legally required. No less important is the role of individual action to exercise rights proposed in these reforms. Ordinary citizens must have easy means for "resigning" from data systems they are not legally required to participate in. They must be able to act directly to block commercialization of "their own" data and set their own terms for such commercial use. And they must have easy and readily understood means for seeking redress, including punitive damages, for willful violations of these new rights. Without necessarily granting anyone *full* control of data on himself or herself, the reforms aim to make everyone a participant with something important to offer (or withhold from) other parties in these respects.

- Nothing in this program rejects *categorically* the need for carefully calibrated institutional surveillance and personal decision-making based on such monitoring. Attending to others' actions and calibrating one's own actions in response are essential to all social life. We must have a means of knowing that the person who proposes to remove one's appendix is indeed a surgeon; or that the staff at the school where one sends one's children do not have histories of cruelty to children; or that one's favorite restaurant does not have a kitchen harboring the next pandemic virus. But privacy advocates must insist that "needs" for personal information implicit in these relationships are not peremptory—and that the costs of meeting such needs may sometimes not justify the rewards.

- The proposed reforms assume a bright line between personal information held in the form of personal-decision systems and other forms of recorded personal data. The measures proposed do not apply to information recorded for purposes of public discourse—political statements and other journalistic and historical writing, for example, or research for electoral campaigns, or screenplays for films related to public issues, or even the writings of obsessive gossips bent on recording the moral shortcomings of their neighbors. But when use of given bodies of information takes the form of routinized patterns of decision-making based on standard sources of filed data, to address enduring needs of organizations—knowing how much tax people owe, or who "needs" to be subjected to the attention of antiterrorist forces, or which prospective medical patients bring with them high risks of lawsuits—then carefully targeted restraints like those proposed here must come into play.
- The administration of any reformed privacy code requires new institutions—either in the form of a European-style Personal Data Protection Board or within an established regulatory agency like the Federal Communications Commission. Individuals must have maximum resources to resolve complaints against institutional users of their information through direct contact with those institutions. The ability to seek damages under a private right of action—the right to sue—is essential. But there will always be points at which mediation and decision-making by an authoritative outside institution is necessary. Whether the forum is civil court or a regulatory body, the resolution of disputes must contribute to an evolving body of precedent, so that future parties to similar disputes may anticipate the likely outcomes of their own claims.

What results should we expect from enactment of these reforms? What changes will Americans experience, in the wake of their adoption?

First would be a dramatic *net reduction* in the total of written or computerized files held on nearly every American. Privacy advocates should count this as a major step ahead in itself. The first line of defense would be Reform Three, which proscribes collection and retention of personal information, unless the holders can demonstrate that the data in question are "functionally necessary," to invoke Jerry Kang's standard.[24] Similarly, many records held by both public and private organizations are created and retained *unilaterally*—that is, without our knowledge or consent, and without statutory authority. In the absence of informed consent or legal requirement, these vast domains of personal data should be off-limits for any personal decision-making.

A second far-reaching effect would be to minimize the mystery and indeterminacy with which Americans must now regard the origins, maintenance, and use of filed data on themselves. A leitmotif of present-day American experience is astonishment, often merging into exasperation, over the origins of personal data that impinge upon our lives. Why do my Google search results differ from my friend's, when we enter the same search terms? How did the producer of pornographic films get the idea that I prefer the precise tastes that their (unsolicited) advertising is pitching to me? What strategies, based on data from Facebook files, were those at Cambridge Analytica planning to mobilize to target people like me?

Reliance on the portals envisaged in chapter 4 should enable any internet user to track the movements and the uses made of his or her records—and to determine which of these activities are consistent with reformed law in this connection. The self-activating function of the portals should make it possible for anyone to monitor the contents of his or her records—and to determine whether the organization holding the records has legal grounds for doing so. Every organization making decisions based on filed personal data should be prepared to account for the legal status of its uses—ultimately, on pain of civil action from the subject or administrative sanction by a Personal Data Protection Agency or other responsible authority.

Additionally, we should expect the balance of *initiatives* based on recorded personal data, and concomitantly of *power relations,* to become much more symmetrical between institutions and individuals. Under these reforms, there will simply be many fewer openings for organizations of any kind to act on the basis of filed information maintained without the consent or knowledge of those depicted by it. The new requirement for legal bases, the right to resign, the newly created property right over commercialization of data on oneself, and other new protections should leave everyone better informed and more deeply engaged in the uses of his or her own data. We should see fewer junctures at which institutions direct communications urging or demanding action from individuals, and more such pressures in the opposite direction. It should be far easier and more effective to confront the sources of junk communications with challenges to the legal grounds for generating them. And in extremis, the response to such importunities could be legal claims for damages.

Nevertheless, the reforms proposed here leave unanswered some key questions concerning the size and organization of data-holding institutions, and the prospects of ordinary data subjects seeking satisfaction from them. For example, a right to resign from data systems that function essentially as blacklists affords especially effective tools, from the standpoint of aggrieved data subjects. If, as in the case of Margot Miller (chapter 2), the only message conveyed by the listing of one's name is "Danger!," any rational consumer would of course want to resign at once. Here, the absence of a record held by any one data system would simply be consistent with the absence of negative information. By contrast, in a system like today's consumer credit reporting, the buyer of reports can reasonably assume that any available credit information, positive or negative, will be included. Thus, the absence of *any* credit record for an adult consumer would suggest to a skilled reader a very unusual life history (institutionalization? a hippie lifestyle?)—probably a danger sign for many decision processes. In short, having *some* kind of record is indispensable for accessing credit, jobs, and any number of other relationships, so resignation from that system may amount to self-punishment.

In short, if you're trying to escape the effects of negative personal information, it's always better to act in an environment that tolerates a lot of dispersed, incomplete databases. Without the advantages of such haphazard recordkeeping, a single bad report can indeed cause much damage—not just to credit applicants, but to those seeking housing, jobs, or positions of trust. Recall the devastating effects on the life of one Amit Patel, recounted earlier in this chapter. Because his credit report (unbeknown to him) identified him as a terrorist, he found himself blocked not only from access to credit, but also from housing and employment. The extreme centralization of credit reporting in this country, and the relentless determination of the Treasury Department's Office of Foreign Asset Control, essentially shut down his life—until it was revealed that the damning information applied against him really referred to another person with virtually the same name.

America's current systems of credit reporting leave consumers highly vulnerable to a single bad report. Present-day practices allow consumers to add to their credit reports one statement of extenuating circumstances in connection with a disputed account that carries a bad report. But one doubts that credit managers and other decisionmakers grant much weight to these statements. Accordingly, I sometimes wonder whether the interests upheld in this book would be better served if our system of credit reporting offered a kind of "Fifth Amendment" option. This would grant every consumer the right to strike one "black mark" off his or her credit record, say, every X number of years—as a kind of safety net against high-handed treatment from credit grantors. If every consumer had the option of censoring the record of just one occasional falling-out with a creditor, perhaps more credit disputes would be solved by compromise. Under those circumstances, both sides might share a motive to end the relationship with a record that both could live with.

Finally, these reforms must apply to filed personal information in the broadest sense—including the details of interactions between providers of data services and their users. The fact that a given consumer posts his or her "likes" or other commentary on Facebook, or generates a profile of his or her interests via Google, must never be taken to mean that these institutions are at liberty to do as they wish with the details of these exchanges. Not just the original information provided by the subject, but also its second-, third-, and (nth)-order interactions between system and user must remain under the subject's control. This includes the analyses of users' inner psychological dispositions and susceptibilities on all sorts of subjects, from choice of hand lotions to voting inclinations, through the character insights Shoshana Zuboff terms "behavioral surplus." These same protections should provide a legal weapon for ordinary citizens to use in fighting government agencies' habit of reselling personal information collected from citizens under legal requirements. The idea that states can require business owners or property owners to provide basic information on their financial situations to state agencies, and then benefit from resale of such data on the market, should represent violations of the commercial property rights proposed in chapter 6.

Perhaps the most important repercussions of these reforms will be in the *culture* of Americans' attitudes toward their own information. My aim is to change the way Americans think and communicate about the role of personal information as a force in our way of life. Taken together, the reforms should countervail against the pervasive fatalism that leads Americans to ascribe the loss of control over their data to "Technology." Thus, none of these measures is intended as a technological "fix"—that is, a response to troubled situations achieved through alteration of our *tools,* rather than our social relationships. Instead, the overarching aim is to *politicize* the social roles of personal information, in the classic and best sense of that term. Far from aiming to make conflict and controversy over such matters go away, these measures seek to engage us all more fruitfully in defining the kind of informational world we want to inhabit. Debates over the privacy of institutionally held personal data have suffered greatly from being portrayed as beyond the ken of ordinary citizens and consumers—too complex, too arcane, and perhaps too boring. In fact, these debates are essentially about who-can-know-what-about-whom—and who can take what action on such knowledge. These matters need never be boring or excessively complex, once their repercussions on our lives are apparent. And these reforms are intended to make them apparent to everyone.

Consent: Formal and Substantive

Looking back at the full trajectory of these arguments—and especially that of the proposed reforms—I am struck by how often notions of *consent* play a pivotal role. I suppose I should not be surprised. Any work devoted to improving the position of ordinary citizens and consumers vis-à-vis the organizations devoted to tracking, monitoring, or enforcing the rules that bind them will want to increase the number of points where people can grant or withhold their consent.[25]

But consent is a slippery concept. If someone points a gun at me, demanding "Your money or your life!," and I respond by handing over all the cash I have, one would hardly say that I gave my consent to the transaction. When the costs of noncompliance are this great—

for example, renunciation of most other personal values, perhaps including life itself—we hardly think a decision has been made on the legitimate merits of what one is consenting to. In a world like the one we now inhabit, where personal information is sometimes seen as the equivalent of coal as a basic resource during the industrial revolution, demands for it are often overbearing. When asked in an employment interview for consent to the interviewer's request to draw a credit report for use in the considerations, is the candidate exercising authentic consent? When a police officer, during a traffic stop, asks the motorist for consent to examine the contents of the car's trunk, can we interpret the driver's response as a reflection of his or her authentic view of the validity of the request? When any of us "consents" to all the terms and condition of software use, or website access, or membership in a new association of users, is the behavior that follows as we click through the boxes an expression of our balanced weighing of all the alternatives?

In fact, the very success of today's powerful personal-data-management organizations contributes to making consent to the use of such data more problematic. The IRS, for example, stresses the secrecy of its extensive and far-reaching files on the intimate financial affairs of most Americans. But as we all know, people's disclosures to the IRS are not really secret, if only because taxpayers remain free to consent to the sharing of their tax returns—which are of course all the more sought-after by privacy-evading parties because of the social forces and demands surrounding their compilation. The fact that this documentation is created on pain of serious penalties, if the IRS should later determine that the statements were false, only enhances the interests of deciders ranging from college admissions committees to mortgage bankers to parole boards.

The point here is that surveillance institutions often do their work so well as to multiply the incentives to providing consent to the information they have captured. If an outsider to a particular personal-decision system knows with total confidence that particular forms of personal information are compiled in one place, it becomes all but impossible for the subject of that information to deny that it exists. When asked by a mortgage lender for copies of one's recent income tax returns, claims that no such documents exist are unlikely to win the day for any mortgage applicant.

Consent, we might say, may come anywhere along a spectrum from purely formal to substantive. Consenting to sign over one's assets at the point of a gun would be at the *purely formal* end of the spectrum: the decider certainly means what he or she avers, under the circumstances; but the decision-making does not emerge from a wide range of attractive alternatives. Consenting to share one's personal information in a situation where the decider is in a position to weigh meaningful options both in favor of and against disclosing—where the decider can indeed "take it or leave it"—could be bracketed at the *substantive* end of the spectrum.

Perhaps the aim of institution-building in the world of personal-decision systems should be to maximize the proportion of fully substantive decisions people make over their data, and minimize those that are purely formal—that is, involving virtually no choice at all. That is why the reforms proposed here aim broadly at creating relationships where individual citizens and consumers have something desirable to offer the organizations they deal with—and where they can indeed "take it or leave it" in relation to what they seek from those organizations. That much should be evident in Reform Four, which requires consent for the release of information from personal-decision systems not required by law. No less important here are Reform Ten, establishing the right to resign and withdraw from any data system whose attention is not desired, and of course Reform Eleven, establishing property rights over commercial exploitation of anyone's personal data. These new rights for individuals, and responsibilities attributed to organizations, aim at helping place the decision-maker at a point where he or she indeed has meaningful decisions to make.

Could It Happen Here?

But are such sweeping changes even possible?

Some readers will fault these appeals for new directions as coming, so to speak, without a road map. To anticipate a sweeping shift of control over personal information, from the most prosperous and entrenched institutions in our country to the currently atomized and disenfranchised grass roots, they may hold, strains the imagination. What sort of Zen-like guidance is it, after all, that doesn't even tell us what number to dial, when we pick up the phone on day one, to start organizing?

To some of these charges, I plead not guilty. I have tried, in the preceding pages, never to minimize the strength of the forces of opposition to the reforms proposed here. But neither have I sought to prescribe any single strategy or scenario to bring the reforms to reality. This is partly because I do not believe that there is any one route to the realization of these objectives. I can imagine many chains of events that would move Americans to successful mobilization to take privacy seriously, but none that promises certainty. Appendix 2 seeks to convey an array of settings and interests that have triggered urgent public demands for better privacy—as well as false starts and missed opportunities to press such demands.

Instead, my first aim here is to create a persuasive *vision* for a more privacy-friendly world—that is, a sense of how well the values of privacy and pluralist democracy could be served, without requiring categorical renunciation of other basic public values. I want to approach the questions before us afresh—in utopian terms. Utopian not in the sense of letting imagination run wild, but in the sense taking maximum advantage of the new opportunities afforded by available technologies for a rich civic life—after subtracting the effects of power differences, entrenched advantage, and sheer ignorance of alternatives that now shape public deliberation on the optimal role of personal information in our civilization.

Without some such vision of the ultimate possibilities, we risk finding ourselves mired in the particularities of the present, responding eternally to discrete individual threats and insults to privacy—without a comprehensive picture of a better world that might win us allies and form the basis of a broad-based movement for comprehensive change.

Do we have any precedent for such sweeping change? Could a grassroots movement dedicated to reforms like those proposed here ever succeed in getting Americans to upend the burden of inequality in control over personal information that we inherit?

In the spring of 1988, Governor Michael Dukakis, of the liberal state of Massachusetts, was front-runner in the Democratic presidential primary. He was having a bad moment at a hotel ballroom in California on June 7, however, because of his position on adoptions by same-sex couples. "Despite the demands of gay men and lesbians," *Washington Post* reporter David Broder wrote, "Dukakis repeatedly refused to back down from his state's policy of giving preference in foster-child placements to heterosexual couples with other children. 'I just happen to think that's better,' he said."[26]

Dukakis was actually a bit ahead of his time. He had already vetoed legislation in Massachusetts that would have prohibited gay and lesbian couples from being foster parents—a risky move in the 1980s. Some observers felt, in 1988, that even that progressive step may have cost him votes in the presidential election, which he lost to George H. W. Bush. By 2004, Massachusetts became the first U.S. state and the sixth jurisdiction in the world to legalize same-sex marriage. The conservative backlash was intense, with a number of states passing laws outlawing the practice, often to have them quickly struck down by state and federal courts.

President Bill Clinton, known for good relations with the gay and lesbian communities, feared that his perceived sympathy to gay causes would cost him his bid for reelection in 1996. In response, he supported the Republican-sponsored Defense of Marriage Act, which blocked same-sex couples whose marriages were recognized in their home states from receiving federal benefits available to other married couples under federal law. Clinton signed the bill into law in September of that year. He publicly recanted those actions in 2009. By 2014, more than 70 percent of the U.S. population lived in states where same-sex unions had become legal. In June 2015, the Supreme Court struck down all state bans on same-sex marriage, legalizing it for all fifty states. In December 2022, the House of Representatives and the Senate passed, in bipartisan votes, the Respect for Marriage Act, mandating federal recognition of same-sex marriages. President Biden at once signed the Act in a flamboyant White House ceremony.[27]

From 1988 to 2022—thirty-four years, a bit more than a generation. That is a remarkably swift turnaround for what many students of public opinion in Michael Dukakis's heyday would have bracketed as a set-in-stone feature of public morality. As of 2021, Gallup reported public support of same-sex marriage at 70 percent, a record high, up from some 27 percent in 1997.[28] I would regard this as an acceptable timetable for a revolution in Americans' attitudes toward privacy.

Could public attitudes about control over one's own personal information evolve as rapidly? I see no reason why not—though I hardly see such a shift as inevitable. Public opinion polls have often shown strong public acceptance of proposals to grant broader power over filed information on oneself. Such attitudes appear to have fueled the wellsprings of approval for California's 2018 privacy law—which of course began as a ballot initiative. The idea seems intuitively simple: if the economy is evolving so that personal information takes on such value as a resource for organizations of all kinds, why shouldn't ordinary people not share in the control of that resource, when its origins lie in their own lives?

And note one striking point of contrast between same-sex marriage and interracial marriage, on the one hand, versus people's attitudes toward control over "their" data on the other. The former issues are deeply rooted in historical American culture, buttressed by religious conviction and people's conceptions of their own identities. The vitriol evoked by challenges to established practices in both these cases serves to index the strength of support they enjoyed. By contrast, notions of the rightfulness of maintaining control over data on oneself might readily be understood simply as an extension of First Amendment rights of free expression. Freedom not to disclose information about oneself, that is, might be characterized as the obverse of a right to put one's views and experiences forward publicly.

Nothing to Hide?

But "we haven't experienced any privacy disasters thus far," some will respond—unconvincingly, in my view. "Nobody is really interested in my data," others will insist, "or if they are, I don't see what harm can come from their knowing." Or "This country has done so well by encouraging the free flow of information," some may observe, "why put the brakes on personal information, any more than any other form of data?" Or—in a phrase sure to evoke groans from privacy advocates everywhere—"After all, I have nothing to hide."

In fact, having things to hide is part of the universal human condition. Our views of those around us never precisely match their views of themselves. There are always moments when we experience desires or perceptions that could give rise to distress or conflict with those around us—or, on a larger scale, with those in charge of the institutions that form the contexts for our lives. Nobody wants to act on all the potential tensions implied by these differences. In fact, life would be unlivable if we sought to do so. Deeply learned privacy instincts are hence indispensable conditions for all but the most evanescent social ties. Without such sensitivities, we might as well be inhabitants of an ant colony, each of us playing exactly the roles dictated by the position we are born into.

Of course, there are *costs* associated with taking privacy seriously. If we reduce the amount of personal information mobilized for institutional decision-making, even to a moderate degree, we open the way for compounded mistakes—prisoners released who become recidivists; tenants accused of vandalism who receive a second chance yet vandalize again; even terrorists who slips through surveillance nets with their murderous intentions intact. Privacy advocates need to face these risks frankly, rather than insisting that there is no downside to their position. What we are proposing is not a world without defenses against destructive behavior, but one where processes of tracking and discrimination deliberately stop short of their maximum intensity. Because the alternative is worse: a world where temptations to institutional overreach grow in tandem with the art and technologies of surveillance. And one where we as citizens, neighbors, and family members are literally never alone, but constantly share every moment with uninvited visitors.

Ultimately, the most compelling reason for turning away from the path leading to total surveillance is that hardly anyone wants to arrive at the destination. Institutionally, its allure of security and predictability are apt to represent false promises. By creating a world where the authorities are always deeply in charge, we run the risk of living in one where the authorities' failures become disasters for everyone. A pluralistic world, by contrast, grants us the best chances of keeping repressive powers in check, or at least reserved for authentic and certified emergencies. Political repression is an endemic threat in all social life, and a particular hazard for open societies whose citizens believe "it can't happen here." A permanent diet of maximum surveillance is hardly a reasonable strategy to escape such hazards.

Considerations are much the same in our lives as individuals. Does anyone really want to live in a world where we conduct ourselves in all contexts as though we have "nothing to hide"? Intimacy and candor born of mutual confidence are always great achievements, as Charles Fried argued decades ago. But they assume states of less-than-perfect intimacy and candor, tempered by *choices* that offer us something better. We do not want to live in a world that makes no space for such a choice.

Notes

1. Heidi Messer, "Why We Should Stop Fetishizing Privacy," *New York Times*, May 23, 2019.

2. Christopher Evans, *The Micro Millennium* (New York: Viking Press, 1979), 152.

3. See Josh Chin and Liza Lin, *Surveillance State: Inside China's Quest to Launch a New Era of Social Control* (New York: St. Martin's Press, 2022).

4. Adapted from James B. Rule and Han Cheng, "Coronavirus and the Surveillance State," *Dissent*, Summer 2020.

5. Consumer Financial Protection Bureau, "List of Consumer Reporting Companies," 2019, https://www.consumerfinance.gov/consumer-tools/credit-reports-and-scores/consumer-reporting-companies/companies-list/.

6. Francis Mailman Soumilas, P.C., "Falsely Labeled a Terrorist in Credit Reports Cause Chaos [sic]," Oct. 23, 2015, https://www.consumerlawfirm.com/falsely-labeled-terrorist-credit-reports-cause-chaos/.

7. Anti-Terrorism and Effective Death Penalty Act of 1996, Public Law 104–132, *U.S. Statutes at Large* 110 (1996): 1214.

8. Adam Liptak, "At the Supreme Court: A Case of Abuse of the No-Fly List," *New York Times*, Feb. 24, 2020.

9. "*FNU Tanzin v. Tanvir*," *Ballotpedia*, http://ballotpedia.org/FNU_Tanzin_v._Tanvir (accessed Apr. 13, 2023).

10. "DHS Inspector Finds Backscatter Scanners Meet Safety Standards," *Homeland Security Today,* Feb. 29, 2012.

11. Willis Bruce Dowd, "England's National Insurance Act," *The American Federationist* 19 (July 1912): 555–56; Morris Hilquit, Samuel Gompers, and Max Hays, *The Double Edge of Labor's Sword: Discussion and Testimony on Socialism and Trade-Unionism before the Commission on Industrial Relations* (Chicago: Socialist Party, National Office, 1914), 72, 107.

12. Carolyn Puckett, "The Story of the Social Security Number," *Social Security Bulletin* 69, no. 2 (July 2009), https://www.ssa.gov/policy/docs/ssb/v69n2/v69n2p55.html.

13. Margaret Reiter, "Government Efforts to Collect Child Support," Nolo.com, https://www.nolo.com/legal-encyclopedia/government-efforts-collect-child-support.html (accessed October 24, 2023).

14. Alfred E. Kahn, "The Tyranny of Small Decisions: Market Failures, Imperfections, and the Limits of Economics," *Kyklos* 19, no. 1 (Feb. 1966): 23–47, 30.

15. Ibid.

16. Davey Alba, "Facial Recognition Moves into a New Front: Schools," *New York Times,* Feb. 6, 2020.

17. Martin Kaste, "Despite Heightened Fear of School Shootings, It's Not a Growing Epidemic," *NPR,* Mar. 15, 2018.

18. Nicholas Confessore, "The Unlikely Activists Who Took On Silicon Valley—and Won," *New York Times,* Aug. 14, 2018.

19. Ashkan Soltani, personal communication, Jan. 18, 2022.

20. Mary Stone Ross, "The CCPA Needs Clarification," *Privacy Perspectives* (International Association of Privacy Professionals), Mar. 22, 2019.

21. California Consumer Privacy Act, CA AB-375 (2018).

22. Geoffrey A. Fowler, "The Technology 202: Privacy Advocates Battle Each Other over Whether California's Proposition 24 Better Protects Consumers," *Washington Post,* Aug. 4, 2020.

23. Jeffrey Dastin, Chris Kirkham, and Aditya Kalra, "Amazon Wages Secret War on Americans' Privacy, Documents Show," *Reuters,* Nov. 19, 2021.

24. Jerry Kang, "Information Privacy in Cyberspace Transactions," *Stanford Law Review* 50, no. 4 (Apr. 1998): 1193–1294, 1249.

25. My thanks here to Professor Kevin Haggerty of the University of Alberta, Canada. In a review of an early draft of this chapter, he had the acuity to point out the pivotal role of consent in attempts to make sense of privacy—and the diplomacy to convince me to add these remarks on the subject.

26. David S. Broder, "Gay Activists Confront Dukakis," *Washington Post*, May 15, 1988.

27. Annie Karni, "The 12 Republican Senators Who Voted for the Same-Sex Marriage Law," *New York Times*, Dec. 13, 2022.

28. Justin McCarthy, "Record-High 70% in U.S. Support Same-Sex Marriage," *Gallup,* June 8, 2021, https://news.*Gallup*.com/poll/350486/record-high-support-same-sex-marriage.aspx.

8 The Future

If I have succeeded, every reader who has persevered this far will have encountered at least one major assertion to disagree with. Many, I'm sure, will be multiply blessed in this respect. Some will certainly object to my inability to ascribe bona fides to the influential institutions shaping privacy policy in this country—from the GAFAM companies through the consumer credit reporting industry and the FBI. Other readers may recoil at the idea that a right to censor one's own records on a case-by-case basis could actually improve the performance of some surveillance systems. Still others will reject the idea of establishing for everyone a strong property right over commercial exploitation of one's personal data. I hope that these disagreements prove fruitful, as I believe they can be. As I seek to persuade my students, we learn much more by articulating our disagreements than by affirming intellectual accord—at least where, as in matters of privacy, the stakes are high.

But on one key point, nearly all privacy-watchers can agree: the multifaceted and far-reaching changes under discussion here are by no means complete. By any reckoning, we commentators are addressing a complex and ever-unfolding situation. Strictly *technological* developments that can be taken as defining the current age—computerization of ordinary organizations and the rise of the internet—are barely fifty years old. Hardly anyone would suggest that we are near the end of their repercussions in public life. Indeed, we may well be much closer to the beginning. If the changes to follow from the early shocks are as profound as what we have experienced to date, the prospects for the next fifty years, or one hundred, are little short of stunning. What can we reasonably afford to hope for in the phases to come—and what do we have most to fear?

For many enthusiasts, as we've seen, the challenge is not so daunting. From their perspective, our role is simply to anticipate what Technology "needs" or where it is taking us—so that we can have our bags packed and be ready to join the trip. The results thus far have been so exciting and so bountiful, in this view, that we'll be happy to discover where we are when we get there.

But in the assessment of monumental social change, as at the movies, the best seats are not always those nearest the action. In historical perspective, the show may just have begun. The intensified flow of personal data *matters,* after all, because it reinforces or undoes other vital social trends and connections—from the changing character of governments and the reach of all forms of authority to transformed face-to-face relationships. In any setting, who-owes-what-to-whom depends on who-knows-what-about-whom. Before we segue into a world where we answer to whole sets of new interests for the most disparate obligations, we must do everything possible to grasp the full magnitude of the changes in train.

One struggles to identify other social transformations of comparable sweep to those we now seem enmeshed in. Consider the enclosure movement in England—another long-running process that, once complete, left the social landscape all but unrecognizable. From the Middle Ages, residents of the English countryside relied heavily on common lands—natural spaces like fields and pastures belonging to entire towns or villages—as key food sources. During the warm season, these lands were reserved for growing crops for residents' own consumption. Later in the year, anyone could use the fields for their geese, sheep, or other livestock. But as early as the twelfth century, large landowners began to claim more and more of these once-common lands for their own purposes, particularly sheep raising. As markets grew and profits beckoned from mass production of textiles, the enclosure movement claimed the livelihoods of many villagers—whose descendants, no longer able to farm, often became laborers in the factories established by textile producers. By the end of the nineteenth century, common lands in England were virtually gone. They live on here and there in cameo forms like picturesque village greens. But as a source of economic livelihood, the commons are all but extinct.

Could privacy—as a set of standards and practices—be headed to the same destination? Are we devolving toward a world where all our personal data fall automatically under the control of large institutions, regardless of our wishes or intents? Will the right to withhold any useful form of personal information from any interested institution become as anachronistic tomorrow as the claim to pasture one's cattle on an English village green today? Given the transformations in the use of personal information over just the past fifty years, it would be rash to reject such possibilities out of hand.

Much public debate over recent decades has focused on when demands for personal information grow "*too* personal"—or on the need to reach a "balance" between loss of privacy and benefits supposedly only available from privacy-eroding routines. These discussions convey a self-flattering impression that we all carry within ourselves a deep-seated and durable sense of boundaries between the public and the private. Such understandings may give comfort to our convictions of our own autonomy, but they do not withstand critical examination.

On the contrary, I have grown convinced that such internal privacy standards are extremely malleable in the face of demands from the larger social contexts. Most of us are more or less gently socialized, as we grow to adulthood, to modern medical care, with its assorted indignities and intrusions. We learn to expose ourselves in the examination room, figuratively and literally, in ways that we have learned not to do in other settings. But when else do we accept signals to abandon the "normal" restraints in these respects? When the authorities insist that we must, apparently.

Thus, I suppose that I should not have been surprised when, as noted in chapter 7, Americans acquiesced in such large numbers to being photographed effectively in the nude, in order to gain access to air travel. We may all be tempted to believe that our emotions, not to say our ethical sense, set down severe limits to the invasions of privacy that we can countenance. But in fact, the signals we receive from the larger social context seem to shape our instincts in this respect, more than the other way around.

Moreover, our sense of legitimate personal boundaries appears to be constantly in flux. We are growing ever more accustomed to disclosing intimate information, or perhaps simply having such information disclosed *about* us, to distant agencies, institutions, and organizations. We expect organizations that pay us for work during the fiscal year to report that income to the IRS—which, of course, makes it imperative that we report the same amount to the tax authorities as well. Interlocking and cross-checking information flows make up the taken-for-granted reality of our relations with the institutions that shape our lives. We cannot navigate what we consider a normal life without keeping our fences mended with such distant, impersonal institutions that we nevertheless expect to treat us "personally"—that is, with attention to the fine details of our records.

BankAmericard, perhaps the first mass-market credit card and the predecessor of today's Visa, nearly failed when the Bank of America tested their innovative product by sending the cards out indiscriminately—for example, to parties whose names and addresses were taken nearly at random from sources like telephone books. The resulting astronomical rates of delinquencies nearly sank the project, until the bank found ways of allocating the cards only to consumers whose records suggested willingness to pay the charges they incurred.[1] Today, of course, the culture of consumer credit is pervasive in this country, and few consumers need to be convinced that ignoring credit card bills will result in serious sanctions via their credit scores.

J. Edgar Hoover, institution builder par excellence, nurtured the FBI by orchestrating yet another novel source of personal information. In the words of then acting attorney general Laurence Silberman, as quoted in chapter 2, "Hoover had, indeed, tasked his SAC's [special agents in charge] with reporting privately to him any bits of dirt on political figures such as Martin Luther King, or their families. It is also true that Hoover sometimes used that information for subtle blackmail to ensure his and the Bureau's power."[2] Adroit and ruthless use of this very special new form of personal information helped make the FBI the powerful institution it has become. It also enabled Hoover to defy retirement regulations and remain in place as director of that institution for forty-eight years.

By foul means or high-minded ones, organizations of all kinds look out for sources of personal information that can bind people in relations of compliance. This is true of governments (bent on extending the "rule of law"); churches (fighting heresy and backsliding among the faithful); companies (bent on building dedication to company policy among their staffs); and supranational entities, like the European Union (seeking to promote "European identity" and thence compliance with EU law).

It appears that we are living through a long-term, secular trend toward growth of larger social units, and within them more intense surveillance and control based on more intimate personal information. These trends have exceptions, of course. Some very large social units collapse and devolve into collections of smaller units—as in the end of the Soviet Union, for example. And certain intense systems of internal surveillance and control also crumble—as I hope that the former FBI director's private system of research and blackmail has done.

The master dynamic at work here, I believe, is that surveillance *feeds* on itself. The more of it there is, the more there can be. The more that transactions, conversations, instructions, agreements, and the like are computerized and retained, the more opportunities there are for conversion of these data into actionable intelligence by institutions bent on using them for new forms of control. As in the mining industry, this often involves data entrepreneurs' going back through the tailings of used and discarded facts, to extract newly useful personal information that might support some new pattern of influence or enforcement. Something like this occurred in the 1990s, when the credit reporting industry discovered that it could sell its credit scores—a product it had long had "on the shelf"—to insurance companies, for use in setting prices and terms for insurance. The more our lives come to be characterized by use of captured data, the more opportunities there are to create new control processes based on these newly found information flows.

None of this is to suggest that these extensions of surveillance and control are inherently sinister or destructive. For example, it's a very good thing to see international law enforcement coordinate in arresting genuinely dangerous figures who flee across national boundaries. Or to see medicine reach the point where physicians in one country interact directly with specialists in other parts of the world to deliver expert medical care (one form of social control, after all) in real time. We must not cast surveillance and control as some sort of morality play, with the forces of evil always on the side of the controllers.

But neither can we shirk the question of just how far we want the overall *reach* of the systematic tracking of human lives to extend.

Even when the ends in question could hardly be faulted—as in the apprehension of dangerous felons—we need to reckon that institutions and practices created for one purpose often come to serve quite different ones. Any system that maintains contact with virtually all employed persons—Social Security, for example—inevitably becomes an irresistible vehicle for other parties in their enforcement goals—for example, a state's Office of Child Support Enforcement.

Thus, any decision—or decision-by-default—to extend the reach of surveillance into any new setting should trigger soul-searching by planners. They should ask, "What further uses of this capability are we prepared to share responsibility for?" Imagine that law enforcement agencies succeed in creating a comprehensive system of automated reading of vehicle license plates—and instant checking of those readings against lists of citizens and consumers of interest to the authorities. Any such system would ultimately reveal an enormous amount of details about the whereabouts and movements of ordinary citizens. This possibility could lead to the quick arrest of people who almost anyone would consider high-priority offenders—kidnappers, murderers, terrorists bent on further crimes—before they even had the chance to flee. This would amount to a stunning invasion of drivers' privacy, certainly. But it is well worth the loss, some would say, in view of the rewards.

But how far, then, would we be willing to go in extending the list of misdeeds against which the powers of such a system should be marshalled? Should such a system be used—as Social Security is—to identify and track those guilty of absconding on child support orders? What about use in apprehending people who have been identified as underreporting their taxable income? Estimates of such unreported income have been put at nearly one in every six dollars due in federal tax revenues, a major drain on government effectiveness.[3] Or for spotting scofflaws, who accumulate parking tickets without limit, and without any apparent intention of paying? Or to apprehend litterers?

One can imagine many other developments, both in technology and in our understanding of human psychology and interactions, that could pose similar ethical and political challenges. Intrusions of violence into everyday life—seemingly random shootings, for example, or hate-crime attacks on members of minority groups, or street fighting among antagonistic political activists—motivate us to wonder how we might anticipate and forestall such events. What if social scientists and medical researchers found a way of predicting the onset of violent behavior—say, from some combination of facial expression and speech analysis? Such an invention is much farther from present-day reality than license-plate-recognition systems, but it would be madness to suppose that it could never occur. Using a device like this, one might scan a crowd for someone about to explode in an angry outburst. Or one could learn other things in an interview or short conversation that the subject might prefer to keep to himself or herself. Or one might spot the one person in an airport security line who's about to start shooting. The insight emanating from use of such tools could certainly have all sorts of life-affirming uses—for example, in defusing tense situations, or in intervening with persons whose violent inclinations were initially unknown. But are we willing to place this sort of invasive power in the hands of any person or agency?

If I have one distinctive message in response to demands for strategies to "protect privacy" against pressures like those considered here, it is the following: There exists *no natural limit* to the connections that might be established between, on the one hand, things that can be learned about human beings and, on the other, behaviors that some influential party may one day urgently seek to control. As human ingenuity continues to establish associations between even the most intimate personal facts and actions whose repression or redirection is held indispensable, the pressures to demand such information will become intolerable. In short, no personal data is inherently too private to stop others from seeking access to it. And indeed, it is often the most intimate and "personal" of personal information that affords the most efficient levers for bureaucratic control.

The tropism that moves us moderns to study and subdue every threatening and disorderly process that we confront—in the human world as well as the natural—is a leitmotif of our times. Often it has served us well, leading to the avoidance of needless suffering or to early warnings of dangerous phenomena—think of hurricanes, erupting volcanoes, or effective attempts to combat disease. But any thoughtful person can enumerate instances where these instincts have led to overreach—from dam projects that prove ruinous to regional control of agriculture to disastrous efforts to convert young persons with homosexual inclinations to heterosexuality.

Against the determination to surveil and control every manifestation of inconvenient or counterproductive behavior, then, my prescription is for a diet of enriched humility. There is no need to appease the appetites of the consumer credit and insurance industries to penalize every incremental increase in "risk" posed by supposedly less competent subsets of the population. There is no need to retain every scrap of information now recorded on the conduct of tenants, or all the criminal histories of all offenders, or every citation that every driver has received. Instead, we need to be content with dividing target populations into broader, rather than narrower, categories and seeking *less information-intensive* alternatives to the most rigorous methods of screening and prediction. Immediate threats to life and limb may well warrant special proactive measures. Elsewhere, we can afford to live with a world that settles for less rigorous discrimination.

Any such strategy goes against some of our deepest cultural grain—above all, our shared thirst for control of the world around us. But the powers of control developed by systems like those considered in this book can lead to mayhem, if turned against those they should be devoted to serving. By reducing the powers they concentrate in the hands of officialdom, we can create room for a more participatory, egalitarian public sphere.

Notes

1. This account of the near-fatal "test drive" of the first BankAmericards was shared with the author years ago by senior executives who had been involved in the experiment.

2. Laurence Silberman, "First Circuit Judicial Conference" (speech, Newport, Rhode Island, June 19–21, 2005).

3. William G. Gale and Aaron Krupkin, "How Big Is the Problem of Tax Evasion?," The Brookings Institution, *Brookings.edu,* Apr. 9, 2019, https://www.brookings.edu/articles/how-big-is-the-problem-of-tax-evasion/.

APPENDIX 1

The Eleven Reforms

Reform One (Chapter 2)

The United States must adopt requirements similar to Europe's, making legal bases indispensable for the operation of all personal-decision systems, private or governmental. Essential to any such a legal basis is the promise of the system to serve a broad range of public interests, and not simply to promote the interests of its original sponsors or any other single party.

Reform Two (Chapter 2)

Renunciation of one's privacy rights must never be a condition for access to products or services, or information about them. Organizations may make the familiar offers of "free" access, in exchange for personal data on the users—provided that the data and its intended uses are correctly specified. But those who make such offers must offer the identical services to all customers for a cash fee, without the capture of personal data, at their "shadow price"—that is, a fee no greater than the market price of providing the service in question. Shadow pricing is a technique economists use to estimate what the price of a given good or service would be, if there indeed were a market for such a good or service. Thus, planners in a country where strawberries were unobtainable in winter might calculate a "shadow price" for strawberries, based on what consumers are known to be willing to pay for other warm-weather fruits that are sold at that time of year. Here, as elsewhere in these proposed reforms, the aim is to preserve for everyone the option of carrying out transactions without loss of privacy.

Reform Three (Chapter 2)

No organization may require personal information from anyone to complete any transaction or provide any service, unless the organization can establish the "functional necessity" of that information in that context. Accounts of such functional necessity must be provided, along with other data on every personal-decision system involved.

Reform Four (Chapter 3)

No data held in any personal-decision system may be used, shared, or disclosed to any other party, except as required by law or with express consent from the subject.

Reform Five (Chapter 3)

Create a corps of expert investigators, endowed with security clearances, legal training, and clear authority to report to some form of national ombudsman-like agency or public prosecutors, to monitor uses of personal information in domains presently off-limits. These investigators must be formally qualified to appear before any court, including the FISA Court, and have resources and staffs commensurate with their responsibilities.

Reform Six (Chapter 4)

Establish a new national privacy-promoting institution, the Master Portals: two closely coordinated websites offering comprehensive information on personal-decision systems maintained by American organizations, both government and private.

Reform Seven (Chapter 4)

The first portal must include the following:

1. The name and business address of the *owners* of the system—or, in the case of government data systems, the agency maintaining the system.
2. The name and contact information of its *keeper*—that is, the party designated as spokesperson for the system, presumably someone available at all business hours.
3. The name of the system (to distinguish it from other personal-decision systems maintained by the same organization).
4. Its *legal bases,* as declared by its keepers.
5. The declared *purposes* of the system.
6. The number of files in the system, the number of persons covered, and the frequency of reporting from the files.
7. Descriptions, using fictional examples, of typical entries in files, including explanations of how to read common codes and abbreviations.
8. An account of typical *uses and destinations* of filed information, for both internal decision-making within the owning organization and the typical outside uses for which data from the system are shared.

Reform Eight (Chapter 4)

The second portal must include the following:

1. The current contents of one's own file, condensed as necessary, accompanied by a legend interpreting symbols, abbreviations, and the like.
2. A brief account of why information held on file is necessary to fulfill the *purposes* ascribed to the recordkeeping.
3. A brief statement specifying the parties outside the organization that may reasonably be expected to receive the filed information, consistent with the *purposes* ascribed to the recordkeeping.
4. A listing of recent entries to and disclosures from the file, including the dates, the identities of the parties forwarding data and those receiving information, and the nature of the material changing hands.

Reform Nine (Chapter 4)

In the public declarations required of every personal-decision system, the declared purposes of the system must refer to purposes shared across a variety of social roles and groups associated with the system. Those purposes must never merely be the purposes of the parties creating the systems.

Reform Ten (Chapter 5)

Anyone who finds herself or himself the subject of unwanted processing of personal data by a personal-decision system operating without legal basis, statutory authorization, or the subject's consent may insist that processing of those data cease, with immediate effect. Repeated failure to comply with subjects' demands along these lines will incur court judgments of compensatory and punitive damages for the operators of the systems.

Reform Eleven (Chapter 6)

No personal information held in a personal-decision system may be sold, rented, "shared," or traded for value for any commercial purpose, except as required by law; by an otherwise valid contract with the subject of the data; or with the latter's freely given consent. Such consent may legally be predicated on payment of some form of royalties to the subject of the data.

APPENDIX 2

International Privacy Affirmations vs. Privacy Setbacks, 1983–2019

To construct the following chart, my students and I cast widely for well-documented instances of two things: first, episodes where ordinary citizens anywhere in the world mobilized themselves for active protest or resistance against what they perceived as privacy-invading laws, practices, or policies ("privacy affirmations"); and second, situations that *might have been expected* to produce protest or demand for strong privacy intervention but failed to do so ("privacy setbacks"). The instances we found of privacy affirmations and privacy setbacks are drawn from a variety of countries; some relate to policies of government agencies, others to actions of private-sector organizations.[1]

The horizontal axis represents the years in which the twenty-seven episodes took place. Each numbered dot represents one case of either a privacy affirmation (upper part of chart) or a privacy setback (lower part). Note that inclusion among the affirmations depends on the strength of the mobilization, rather than the ultimate success of popular demands for change. Even some of the mobilizations involving the largest numbers of participants did not ultimately lead to satisfaction of the participants' demands.

Descriptions of each event, with our reasons for classifying it as we did, appear in the numbered glosses below. For example, the first

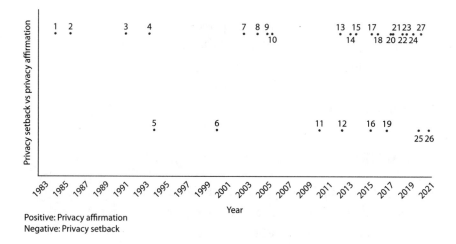

Positive: Privacy affirmation
Negative: Privacy setback

data point, case 1, marks the unexpected outburst of grassroots protest against the scheduled German census of 1983, which triggered mass protest against its perceived intrusiveness. By contrast, case 25 represents the success of Spotify's effort, through its music family plan, to require GPS location data from its subscribers, which we judged a privacy-eroding innovation that might have provoked wide protest but did not.[2]

1. AFFIRMATION: West German Census Revolt (December 1983)

In 1983, the federal government of Germany was scheduled to conduct a general population census. Under the Federal Census Act of 1983, every family was to complete a detailed questionnaire on matters ranging from their living conditions to education and leisure activities. Unexpectedly, these demands triggered strong protest from groups including left-wing activists, consumer protection groups, civil libertarians, and others. Many called for civil disobedience; media coverage was intense. One poll put Germans' disapproval of the content of census questions at 40 percent of respondents. The protesters filed a complaint with the Federal Supreme Constitutional Court—which, to widespread surprise, they won. The immediate result was to reduce some 40 million questionnaire forms to a heap of worthless waste paper.

2. AFFIRMATION: Australian ID Card (June 1985)

In 1985, Australia's Labor government proposed to introduce a universal national ID card aimed at curtailing tax avoidance and welfare fraud. Much to the surprise of nearly everyone, a small protest by privacy advocates and a few others rapidly ballooned into a major series of demonstrations against the card, including one in the city of Perth estimated at 20,000 participants. An array of civil society forces—including small business associations, student groups, and others—joined the opposition. Antagonism to government snooping into citizens' lives was a major theme in the opposition. The government withdrew the proposal in 1987.

3. AFFIRMATION: Lotus MarketPlace (April 1990)

In 1990, the software company Lotus Development Corporation and the credit reporting giant Equifax published a CD containing data on some seven million businesses and 120 million individual Americans, including personal information on the latter such as age, gender, and income. The CD was intended for use in marketing by companies that sought to target advertising directly to promising customers. This product stirred vehement objections among many privacy-oriented people, especially in the tech industries. After a concerted, bottom-up rebellion involving more than 30,000 people, the companies withdrew the product.

4. AFFIRMATION: Clipper Chip (April 1993)

In 1993, the National Security Agency (NSA) proposed a new federal encryption program for email communications that was to be required for all foreign and domestic email. Under this program, the NSA would have sole access to the encryption algorithm, or key, and so be able to decode any coded communication sent within the United States, or sent from this country abroad. Privacy and free speech advocates organized online discussions, news groups, and internet-based petition drives. Within a month, one activist group, Computer Professionals for Social Responsibility, reported receiving some 47,000 appeals to stop the program, which the government ultimately dropped.

5. SETBACK: 407 ETR Toll Road (September 1993)

In 1993, the government of Ontario, Canada, proposed to open an extensive new toll road across the province. Noting that the toll system proposed for the highway could also generate information on where users had traveled, and when, a number of privacy activists insisted on better privacy protections for the personal data held by the system. These privacy-motivated demands garnered very little grassroots support. While drivers now have the option to use an anonymous account when using the highway, few choose this option because planners added a $250 deposit as a requirement for such anonymous use. This poses a barrier to taking advantage of the anonymity option and causes us to rate this a qualified setback.

6. SETBACK: 2000 UK RIPA (February 2000)

In 2000, the Labor government of the United Kingdom introduced and ultimately passed legislation (the Regulation of Investigatory

Powers Act, or RIPA) that accords very broad powers of covert surveillance to government agencies. The announced purposes included enabling police to investigate suspected terrorists, industrial espionage, money laundering, and a variety of other offenses. Among other things, it allows the police, intelligence services, HM Revenue and Customs, and several hundred other public bodies, including local authorities and a wide range of regulators, to demand telephone, internet, and postal service providers to hand over detailed communications records on individual users. This can include name and address, phone calls made and received, source and destination of emails, internet browsing information, and mobile phone positioning data that record a user's location. These powers are self-authorized by the body concerned, with no external or judicial oversight. The proposal set off protests from privacy and civil liberties groups across the political spectrum, but ultimately very little grassroots mobilization from the British public; this was a clear setback for privacy interests.

7. AFFIRMATION: Passage of North Dakota Financial Privacy Law (June 2002)

In 2002, a grassroots group of North Dakota voters collected some 17,000 signatures to oppose laws enabling banks to disclose clients' financial and personal information to outside parties. The activists proposed to change to an opt-in system from the prevailing, and much weaker, opt-out system, where silence was viewed as consent. Overwhelmingly choosing to affirm financial privacy, 87,446 people voted to veto the opt-out standard.

8. AFFIRMATION: Do Not Call Line (October 2003)

For years, unwanted robocalls (junk phone calls) to Americans' residences were a major source of complaint to the Federal Trade Commission. In 2003, the FTC began accepting self-nominations for the national Do Not Call List, which now has 230 million numbers. The list is very imperfect, particularly given that many robocallers conduct their operations from outside the United States. Over 19,000 Americans continue to phone complaints to the FTC every day, even as the Do Not Call List remains in place. Despite the failure of the list to achieve the goals held out for it, this case represents a clear privacy affirmation—evidence of ample public mobilization on a clear-cut privacy issue.

9. AFFIRMATION: Defeat of Labor Party proposal for a National ID Card in the UK (December 2004)

In 2004, Great Britain's Labor Party Home Secretary David Blunkett put forward a plan for a biometric identification database that would include a file on every adult legal resident in the UK. Each resident would be issued a unique number and card linked to their data. These documents would be linked to a National Identity Register, and logs would be created and stored documenting the card's use. The NO2ID advocacy organization was formed to oppose the database, and according to the UK Home Office, 30,000-40,000 people participated in NO2ID's "Renew for Freedom" campaign to protest ID registration. Research from coalition partners of NO2ID suggests that the campaign reached around 7-9 million people, and public support in favor of the ID cards dropped from 80 to 50 percent. Following the defeat of the Labor Party, the newly appointed Conservative government decided against the program in 2010.

10. AFFIRMATION: Sutter, California, RFID Case (January 2005)

In 2005, the elementary school in the country town of Sutter, California (population roughly 3,000), was the setting for an experiment in tracking of its pupils via RFID badges. Pupils were supposed to don the badges (held by lanyards around their necks) on entering the campus at the beginning of the day and keep them on until leaving. The badges made it possible for the principal's office to locate every pupil at any moment in the day—making attendance-taking virtually automatic. Many parents objected to the idea on principle, responding with slogans like "Our children are not inventory!" Finally, at a "packed special school district meeting," the company sponsoring the experiment announced that it would end the program in response to the protests. Though the population of Sutter is not large, we rated this a privacy affirmation, given the explicitness of the negative message sent by parents.

11. SETBACK: Backscatter and Millimeter Wave Machines (December 2009)

Early in the 2000s, companies specializing in airport security began to tout the capacities of machines that could produce unclothed images of air travelers without their having to undress. Using technologies akin to those in X-ray machines, "backscatter" and kindred devices seemed to offer the ultimate check for concealed weapons, while speeding people through security lines. The plans met with widespread criticism on privacy grounds and those of health, given uncertainties regarding the long-term effects of the scans on passengers' bodily tissues. But acceptance of the process received a major boost on December 25, 2009, when a would-be terrorist partially detonated a bomb concealed in his underwear on a flight about to arrive in Detroit. Since then, despite objections from privacy groups and other NGOs, public opinion has been acquiescent. This was a clear setback for privacy.

12. SETBACK: Google Screenwise (February 2012)

In 2012, Google unveiled a new program, Screenwise, to track internet use on select home networks. To participate, interested parties were to sign up with their email address on the Screenwise website. Google promised to reward each participant with a $5 Amazon gift card and another $5 gift card for every three months they remained with the program. By installing a small black box in their homes, Google could track 100 percent of a household's web use. Google expressed its intention to share the data with third parties such as academic institutions, advertisers, publishers, and programming networks. Despite critical comments by privacy advocates, no serious grassroots opposition to Screenwise ever manifested.

13. AFFIRMATION: Australia's C-30 Bill on Warrantless Data Access (February 2012)

In 2012, the Canadian federal government attempted to pass legislation that would minimize or abolish the need for a warrant to obtain subscriber information. Although subscriber information was being given out on a voluntary basis by telecom companies, there was an outcry of opinion after Canada's public safety minister, Vic Toews, sponsored "Bill C-30: Lawful Access Act," which would mandate internet service providers to turn over customer information when requested by the police. Citizens started a Twitter campaign against Toews, and a number of petitions were started to pressure the government to kill the bill. Even though the bill was passed, a Supreme Court decision later ruled in alignment with privacy protection.

14. AFFIRMATION: Google Glass (February 2013)

Google Glass was a wearable camera device created by Google in 2013. The product triggered much privacy concern, as the device appeared to record everything that the user saw and heard. Certain businesses, such as Lost Lake Café in Seattle, banned customers from wearing the device on its premises in 2013. Other advocacy groups arose, such as Stop the Cyborgs, that expressed concerns with the surveillance capabilities of the device. The Stop the Cyborgs campaign had 3,352 followers at its height. Glass "shut down" in 2015 but has recently relaunched to produce an "Enterprise Edition" for use in factories, with "Glass Partners" across the world.

15. AFFIRMATION: Care.data in the UK (April 2013)

In 2013, the UK government launched NHS care.data, a program designed to integrate patient health and social benefits onto one platform. NHS care.data was also intended to supply researchers with population health and social information. Many citizens became concerned about what they considered its sharing of sensitive medical information with commercial companies without the explicit consent of the patients. "The beleaguered scheme faced almost relentless criticism since it was first announced three years ago. Concerns centered on the sharing of sensitive medical information with commercial companies without the explicit consent of patients. More than one million people opted out of the scheme" (*Wired*, July 6, 2016, www.wired.co.uk/article/care-data-nhs-england-closed). In 2016, as a result, the program was officially closed. In view of the role of public objections in this episode, we classify it as a privacy affirmation.

16. SETBACK: Smart Meters in the United States (December 2014)

Since 2006, Pacific Gas & Electric (PG&E) has installed nearly 9.1 million smart meters across Northern and Central California. The consumer can opt out for an initial $75 setup charge and a $10 monthly charge thereafter. Opt-out fees were imposed in California in December 2014. Promptly after these opt-out fees were levied, a new wave of small protests occurred because conservatives and civil libertarians viewed the monitoring of home appliances via the new appliances as a breach of privacy. But this short-lived wave of protests never developed into a larger movement, and smart metering still prevails in the PG&E service area. We bracket this as a privacy setback.

17. AFFIRMATION: Australian Mandatory Metadata Retention (March 2015)

In August 2014, the government announced its intention to update Australia's telecommunication interception laws. This was part of broader efforts to enhance powers available to security agencies "to combat home-grown terrorism and Australians who participate in terrorist activities overseas." This included developing a mandatory "metadata retention system." This mandatory data retention scheme would require that telecommunication companies and internet service providers retain records of citizens' telephone and internet communications for two years. In 2014, hundreds of thousands of Australians gathered at the Parliament House in Canberra to protest the passage of these new telecommunication interception laws. Despite the public backlash and protest, legislation to introduce a mandatory telecommunications data retention scheme was passed by both Houses of Parliament in March 2015, and this metadata retention system continues to infringe on the privacy rights of Australian citizens daily.

18. AFFIRMATION: Suspension of Thailand's "Single Gateway" Internet Proposal (August 2015)

In Thailand in 2015, the ruling military junta introduced a measure aimed at controlling the flow of information into the country via the internet. Called the Single Gateway proposal, this measure was widely held to be an attempt to create a firewall to deflect information and communications useful to opponents of the regime. Many civil society organizations joined in the resistance, including groups representing internet users, civil rights organizations, environmental activists, online gamers, and others. Faced with this resistance, the government withdrew the proposal in October 2015. Some of the protesters feared that the government could revive the project, but because of the broad-based resistance, we bracketed this as a clear success.

19. SETBACK: India's Aadhaar Electronic Payment and Population Registration Project (March 2016)

Legislation adopted in 2016 authorized the Modi government to proceed with implementation of this national system affording payments of virtually any debt to any registered party, at minimal cost. Today nearly all Indian citizens rely on the system, usually accessed by cell phone, to settle charges ranging from rickshaw rides to rent payments. But the system requires all participants to register, providing biometric ID and a variety of other personal information, which the government in turn uses, among other purposes, for enforcement of tax obligations. Many computing specialists and civil libertarians have argued against the system in its present form, citing its obvious potential for tracking and threatening citizens regarded as political enemies by their government—particularly given the increasingly authoritarian directions of the Modi administration. But the widespread acceptance of the program and its apparent economic benefits have thus far eclipsed these anxieties.

20. AFFIRMATION: 2016 Australian Census (July 2016)

In 2016, Australians became concerned about the security of the personal data they were required to share with the government in the quinquennial national census. This was Australia's first "digital census," meaning that all citizens were expected to fill out census forms and questions online through a government-hosted website. The Australian government revealed that the census website's security used an algorithm to help protect collected data against hacks and cyber attacks, but Australian citizens pointed out that the majority of internet browsers no longer utilize the algorithm. Nearly 7,000 people boycotted the census in 2016 due to privacy concerns, and close to 100,000 expressed their worries on social media platforms. Despite national distress over the security of personal data, Australia ignored public pressure to develop and utilize a new algorithm and threatened to fine citizens who refused to participate in the census.

21. AFFIRMATION: Expansion of Thailand's Computer Crime Act (December 2016)

In 2007, Thailand implemented a new Computer-Related Crime Act, which was originally intended to offer citizens protection against fraud and data breaches. After its enactment, a number of amendments have encroached on citizens' privacy rights. In 2016, an amendment threatened to give overly broad powers to the government to "restrict free speech, enforce surveillance and censorship, and retaliate against activists" (Human Rights Watch, 2016, www.hrw.org/news/2016/12/21/thailand-cyber-crime-act-tightens-internet-control). Thai citizens considered this an assault on their privacy rights and garnered nearly 360,000 signatures on a petition against the amendment before the national legislature's vote. Domestic activists also started online protests, and a number of international rights organizations released statements against the amendment. Despite concerns expressed by civil society, business, and diplomatic representatives, this controversial amendment was unanimously adopted. In view of the extent of the mobilization, however, we bracket this as a privacy affirmation.

22. AFFIRMATION: Opposition to the Repeal of the FCC's 2016 Broadband Privacy Order (April 2017)

In 2016, the Federal Communications Commission (FCC) instituted an order requiring internet telecommunications carriers to protect the confidentiality of consumer information. A key implication of this order was that one's ISP or telephone carrier could not sell, trade, or otherwise disclose one's identity or any information about one's telecommunications use to outside parties. This measure, passed with Democratic Party support in the House of Representatives, was opposed the following year, when the body was controlled by Republicans. With support from telecom companies (e.g., Verizon and Comcast), tech companies (Facebook and Google), and other business interests, a repeal of the measure was passed in 2017 and signed by President Trump. Nevertheless, grassroots opposition to the overtly anti-privacy repeal measure garnered considerable support, including a statement bearing 100,000 signatures sent to Senate Majority Leader Mitch McConnell. Despite the net defeat for privacy, this mobilization of public support warrants including this episode as an affirmation.

23. AFFIRMATION: Alibaba/Ant Financial
Customer Privacy (January 2018)

In 2004, China's retail giant Alibaba partnered with Ant Financial to develop a third-party mobile and online payment platform called Alipay. The Alipay app provides features such as credit card bills payment, bank account management, insurance selection, and digital identification document storage. But in 2018 it was discovered that Alipay's mobile-payment platform did not properly notify people that enrolling in their credit-scoring system would "give Ant sweeping rights to their personal financial data, allowing it to be shared with other companies and third parties" (*Wall Street Journal*, January 10, 2018, www.wsj.com/articles/china-swats-jack-mas-ant-over-customer-privacy-1515581339). A critical post on this matter by Yue Shenshan, a lawyer and legal commentator for state media, quickly garnered more than 22,000 likes and almost 9,000 comments after five days. Following these expressions of concern, the Cyberspace Administration of China (China's internet regulator) ordered Ant to fix the problem.

24. AFFIRMATION: California Consumer
Privacy Act (June 2018)

In 2018, a small group of activists led by Alasdair Mactaggart launched legislation that would become California's landmark Consumer Privacy Act. The original petition for a ballot measure featuring Mactaggart's measure received over 600,000 signatures. Among other things, the resulting Act, modeled after the ballot measure, granted consumers the right to prevent businesses from selling their data and imposed new penalties on businesses that don't comply. It was signed by Governor Jerry Brown in 2018 and took effect in 2020.

25. SETBACK: Failure of Public Mobilization against Spotify's Efforts to Require Location Disclosure (September 2018)

In 2018, the music service Spotify sought to impose a requirement on its family-plan subscribers to verify their location or risk losing access to its music streaming service. Apparently the company suspected that some customers had been sharing family plans with unrelated individuals, in order to pay less. Customers were uncomfortable with these demands. Instead of entering a mailing address for the main account, for instance, the company asked for customers' exact GPS location. The company's emails also threatened that failure to provide the new information could cause customers to lose access to the service. Still, little public resistance was reported.

26. SETBACK: Australia's Trusted Digital Identity Framework (June 2019)

In 2017, Australia's Digital Transformation Agency released the first public draft of its Trusted Digital Identity Framework. This framework laid the foundation for extended development of myGovID and Digital ID, two apps that were created to provide streamlined citizen identity verification across a range of government and private-sector services such as social welfare, veterans' affairs, health care, and online banking. Both apps were released without fanfare or large-scale marketing campaigns to lure users. Since the unveiling of the app in 2019, more than 7,000 Australian citizens have created a digital identity using the platform. A number of security researchers have advised Australian citizens not to download or use myGovID due to a critical design flaw that would enable hackers to access user information and controls quite easily. The Australian government responded by stating that it did not intend to fix the issue because it did not prove to be a valid vulnerability. Despite the security concerns raised by researchers investigating these apps, and in sharp contrast to the popular revolt against the Australia Card proposal in 1985, no serious opposition to myGovID or Digital ID has been recorded.

27. AFFIRMATION: China Facial Recognition Case (November 2019)

In 2019, a Chinese university professor sued the Hangzhou Safari Park for "violating consumer protection law by compulsorily collecting visitors' individual characteristics" after it made facial recognition registration mandatory for visitor entrance. Following this case, the hashtag ChinasFirstFacialRecognitionCase was created and used on Weibo; a thread about the suit garnered more than 100 million views on the platform, and as many as 10,000 users expressed concern about the technology. Since the lawsuit and public outcry, the park has compromised by offering visitors a choice between using the previous fingerprint system and high-tech facial recognition.

Notes

1. This appendix was produced by Rachna Mandalam with the collaboration of other URAP students, including Angelica Vohland. James Rule and the student collaborators contacted writers and researchers on privacy from around the world, asking them to nominate cases for inclusion. After considerable editing and fact-checking, nominations were narrowed to the cases shown here. Additional research was carried out to complete the explanatory glosses and to classify the cases as affirmations or setbacks.

2. Note that the chart entails no claims of statistical representativeness. The cases presented are simply ones that came to our attention, after inquiries to privacy-watchers in a number of countries. The fact that the data-points consist of more affirmations than setbacks, for example, tells us nothing about the frequency of the two sorts of events in any larger population. Instead, the chart aims to illustrate the variety of examples in both categories.

Index

ACLU, American Civil Liberties Union, 49
Acxiom Company, 41, 49, 76
"adding privacy," as reformers' task, 7
Algorithmic Justice League, founding of, 183
algorithms for personal decision systems, in allocating insurance coverage, 172
American Law, its lack of a single theory of privacy, 6
Arney, Dick, 238
Arroyo, Carmen, her efforts to include her handicapped son on her lease, 21–24
asymmetry of position between subjects of surveillance systems and those who direct them, 132
AT & T corporation, 202
attitudes of Americans toward surveillance, subject to changes with changes in social framing, 229
Australian Privacy Charter, 89
autonomy, privacy as, 8

"backscatter" machine for producing images of air travelers without their clothing, 228
Ban the Box movement, 112–119
BankAmericard (predecessor of VISA), near-failure when first introduced in a California field trial, 264
Buolamwini, Joy, 183

Cambridge Analytica, site of privacy-related scandal in 2016 Presidential election, 4
CCPA (California Consumer Privacy Act), passed by the State Legislature in 2018, 240
Canadian Standards Association, 4
cell phone use, a major tool for tracking users, 223
censorship of data on one's self, difficulties of, 2
Center for Democracy and Technology, Washington, D.C., 49
central bank of France, maintains nationwide listing of delinquent credit accounts, 188

centralization of national surveillance systems, in many ways more marked in the United States than China, 227

Chen, Brian X, *New York Times* columnist, reports dismay at the amount of information on himself he discovered in his archived Facebook pages, 93, 106

Chesterton, G. K., quotation, 59

China: surveillance of Covid-19 population, 223; compared to surveillance systems in the United States today, 224

Clearview, software for facial recognition, 32

cognitive overload, imposed by many personal decision systems on their subjects, 133

compartmentalization of personal data held in file, need for implicit in *Records, Computers and the Rights of Citizens* (1973), 90

civil rights movement, 73

Confessore, Nicholas, on the history of California's Consumer Privacy Act, 235

consent: should be required for release of one's property rights over commercialization of one's own personal data, 207; formal and substantive, 250

consent from data-subjects, as a potential basis for cooperation between subjects and management of personal decision systems, 79

consumer credit reporting system, American, has made itself "the only game in town," 168

consumer credit reporting industry: absent in France, 188–190; originated in the United States in the late nineteenth century, 185; right of individual consumers to opt out, 169

consumer credit scores, widely marketed to insurance companies for use in evaluating insurance applications, 100–101

Coret Logic Rental Property Solutions, role of in Carmen Arroyo case, 22

Cotto, Tony, 174–179

court orders and subpoenas to obtain personal information, should carry an obligation to notify the targets of such investigations once the inquiries are complete, 86–87

credit reporting industry, role of in tracking Americans' lives, 225

data brokerage industry in United States, 41

Department of Homeland Security, ability to track movements of domestic U.S. air and rail travelers, 29

discriminating decision-making, the normal aim of surveillance, 30

Dukakis, Michael, presidential campaign, spring 1988, 254–255

Dutch population records, exploited by Nazi occupation forces to hunt down and deport Dutch Jews to concentration camps, 72

Electronic Frontier Foundation, San Francisco, 50

enclosure movement in England, as a model for the scope of changes following from computerization of everyday life, 262

EPIC (Electronic Privacy Information Center), 49
Equal Credit Opportunity Act of 1974, 191–192
Equifax corporation, 199
essentially contested concepts (W. B. Gallie), privacy as an example, 9–10
Evans, Christopher, 220
Exact Data Company, 76
Experian corporation, 197

facial recognition software, 181; banned in a number of U.S. cities, 183
FIPs (Fair Information Practices), role of in Privacy Act of 1974, xv–xvi, 83
FBI (Federal Bureau of Investigation), 265–266; recent efforts of to exempt parts of its national identification system from coverage by the Privacy Act of 1974, 86; maintained files of scandalous personal data on public figures at the initiative of J. Edgar Hoover, 81
feasibility of researching contending programs for taking privacy seriously, 54
fiduciary rule for those who manage personal data, 5, 92
Fifth Amendment for credit reporting, allowing consumers limited options to suppress certain "bad" information from their records, 248
"finders-keepers" prevailing principle governing much of America's use of personal information, 13
"First Amendment fundamentalism," 6
Fourth Amendment to the U.S. Constitution, source of protections against loss of privacy in one's home, 6
France, the country's central bank maintains a comprehensive listing of private citizens' credit accounts that are classified as delinquent, 188
functional necessity of personal decision systems, proposal to require evidence of, 77

Garvie, Clare, Georgetown University, 182
Gebru, Timnit, 183
General Data Protection Regulation (GDPR) of the European Union, 38
geofencing, 221
Gompers, Samuel, 52
government collection of personal information, 106–108
government disclosure of personal data: in criminal records, 112–114

harm: suffered by Brian X. Chen of the *New York Times* from Facebook disclosures of his private information, 93; need to demonstrate in order to establish a breach of privacy, 236
HIPPA (Health Insurance Portability and Accountability Act), 94
Hobbes, Thomas, 222
Hoofnagle, Chris Jay, 44
Hoover, J. Edgar, 265

Igo, Sarah, on American attitudes toward personal decision systems in the 1960s, 82
income projected from sale of permission for commercial use of personal information, 214
individual good, privacy as, 12
Info USA, a data brokerage, 42
information privacy, defined, 8
Information Society, 66

Innovation, property rights over commercial exploitation of personal information as an obstacle to, 208
Institutions for brokerage of rights over commercial exploitation of personal data, 211; as models ASCAP and BMI, 213
Insurance surveillance, 2
IRS (Internal Revenue Service), 2; role of in creating demand for personal information held under its control, 251

Kahn, Alfred E., author of "The Tyranny of Small Decisions," 229
Kang, Jerry, 77
Kant, Immanuel, key inspiration for much European legal and philosophical thought on privacy, 3
Kim [pseudonym] arrested at a gay bar and blackmailed on the threat of exposure of mugshots taken after the arrest, 162
King, Martin Luther, Jr., target of FBI harassment, 73

legal bases: need for U.S. law to require for personal decision systems, 74; should be required for government, as well as private-sector systems, 74
less information-intensive alternatives to today's personal decision systems essential for meaningful privacy protection, 269
Libert, Tim, author of a large-scale study of appropriation of personal information from webpages, 98
"lost in a crowd" experience, 68

Mactaggart, Alaisdair, launched movement resulting in the California Consumer Privacy Act (2018), 237
mass surveillance systems, early forms in Prussia, the UK, and the United States predate computerization, xiii
Mayfield, Brandon, case, 122–127
McCarthy period in the United States, 73
medical records, Veterans Administration, 62–63
Messer, Heidi, 202, 219
multivariate analysis, as a possible technique for setting insurance costs, 175

"needs" of organizations for personal information, 53, 187
New York City Housing Court, 64
NGI (Next Generation Identification System), FBI's effort to exempt this system from the Privacy Act of 1974, 86
"no-knock" raids on private dwellings, 160–162, 226–227
Norris, Onree, target of "no-knock" invasion of his home, 160
NSA's ability to track Americans' telecommunications, 29

OECD (Organisation for Economic Co-operation and Development), 89
organizations, role in sponsorship of surveillance, 1

Palmer Raids, 73
Paramount Marketing Consultants, Inc., 75
Patel, Amit, 248; wrongly harassed as a terrorist 225

peace movement, U.S., 73
personal data brokerages, 76
personal decision systems: interactive, 75; unilateral, 75, 164
personal information: now the equivalent of coal during the industrial revolution, 17; prices of personal information on various groups of Americans, 195, 196, 198–201
Pew Research Center, studies of American privacy attitudes, 44
politicization of privacy, a key aim of this book, 12–15, 85
Portal I, provides information on the declared contents of all personal decision systems, 136
Portal II, provides access to all of one's own files in personal decision systems, 136
"pragmatic" consumer attitudes toward organizations, 44
prescriptions, surveillance over, 2
Presley, Elvis, "one of America's top dead celebrities," 202
privacy: as an anachronistic value, 219; as a new right, defined in EU law, 38
Privacy Act of 1974, xvi, 6, 82–87, 119–126
privacy codes: global rise, xiv; weakness of America's, in relation to privacy-eroding forces, 7
privacy paradox, 45
Privacy Rights Clearing House, 50
privacy-rights-free zone, the United States as, 40
Proposition 24, California's revision of the CCPA, passed in the election of November 2020, 243

prospective organization of a future data-rights industry, 212–215
purposes of institutions, ambiguities of, 149–154

qualified immunity, doctrine that protects police and other public employees against lawsuits for actions carried out in their work, x

racial discrimination in current prices charged for automotive and homeowners' insurance: denied by the insurance industry, 173; asserted by researchers, 173
Records, Computers, and the Rights of Citizens (1973), xiv–xv, 5, 6, 89, 103, 128–129
"redlining" of personal data on planners' maps to exclude Black Americans, 172
reforms proposed in this book as oriented to relatively long-term goals, 16
retinal complexity, a possible correlate in setting insurance rates, 175–176
right to resign from any personal decision system not required by law, 169–170
risk, in insurance: price of coverage based on amount of risk assumed by insuror, 176; based on buyer's perceived responsibility for loss, 179

sale of one's property rights over commercial exploitation of data on one's self should be limited to short periods, 210
same-sex marriage, its rapid growth in acceptance demonstrates that

same-sex marriage *(continued)*
  sharp revision of public opinion on salient public policy issues is indeed possible, 255
"secondary release" of personal data, without permission from the subject, is proscribed in the EU, 48
"sectoral" privacy legislation in the United States, 40
shadow pricing, 77
Silberman, Laurence, decries J. Edgar Hoover's secret files of scandalous data on American public figures, 81
Simitis, Spiros, 73
skin tone affects accuracy of facial recognition software, 183–84
Snowden, Edward, released in June 2013 a vast array of evidence of U.S. government surveillance over U.S. nationals, 236
Social Security surveillance system on account-holders, uses of for non-Social Security purposes, 71
Soltani, Ashkan, 237
Sprint corporation, 202
Stingray, software for tracking cellphone whereabouts, 220–221
surveillance: politics in Lockport, New York, 231–235; potential benefits of intensified, 221; reinforces itself, 266
surveillance systems, how they may prove too effective to be desirable, 71–73
Sutter, California, 281
synergistic value of reforms proposed in this book, 60

Tanvir, Muhammed, threatened by the FBI with inclusion on this country's No Fly listing, 225–227
taste, privacy-related values as a form of, 11
taxation, as source of demand for surveillance, 1
Taylor, Breonna, 160–61
technology: as an independent force undermining privacy, 29; inadequate by itself as explanation for erosion of privacy, xii
"tenants blacklist," New York City, 64
T-Mobile Company, 202
Ton-That, Hoan, inventor of Clearview facial recognition, 33
totalitarianism: the ultimate disaster for values underlying this book, xi
Trans-Union Corporation, 197
trash cans, intelligent, 67
Turow, Joseph, 45

Urban, Jennifer, 44
utilitarianism, doctrine particularly influential in American privacy thinking, 134–135
utopian thinking, virtues of, 14–15

value of data on Americans' whereabouts, 67
Verizon Communications, 202
voter registration, some public officials claim that they lack adequate personal information to register all those who would like to vote, 155
VPN networks, many of these appear to capture much of the personal information entrusted to them, 191

Zuboff, Shoshana, on insights gleaned from big personal data systems on what people are apt to do, or want to do, next, 34, 249

Founded in 1893,
UNIVERSITY OF CALIFORNIA PRESS
publishes bold, progressive books and journals
on topics in the arts, humanities, social sciences,
and natural sciences—with a focus on social
justice issues—that inspire thought and action
among readers worldwide.

The UC PRESS FOUNDATION
raises funds to uphold the press's vital role
as an independent, nonprofit publisher, and
receives philanthropic support from a wide
range of individuals and institutions—and from
committed readers like you. To learn more, visit
ucpress.edu/supportus.